Unit 3

The Solar System

UNIT 3
The Solar System

By middle school, many students have memorized the names of the planets. Some students have not and may never do so. The activities in this unit are for both categories of students. Students will learn details about the excitingly diverse objects in the Solar System, and even more importantly, they will learn key concepts about the Solar System such as: how it is organized, what categories of objects it contains, and how its components move. Students also learn that new knowledge and objects are continually being discovered, causing astronomers to constantly revise their ideas about the organization of the system.

Models

A pervading theme of this unit is the use and application of models. Students begin by evaluating a diagram model of the Solar System based solely on their previous knowledge. Subsequent activities explore a wide variety of other models, such as a participatory moving model of the Solar System and a real-life model of the system of Jupiter and its moons. Students discover that each model depicts some, but not all, of the characteristics of the Solar System. Each model offers its unique insight into one aspect of the Solar System.

To explore the diversity of individual objects in the Solar System, students create unique and fact-filled learning stations and travel brochures profiling an assigned object. After completing their assignments, students "tour" the Solar System by visiting their classmates' learning stations. Through researching objects, preparing learning stations, and touring them, students learn much about the characteristics that define the various objects in the Solar System.

Students have long been fascinated by "Planet X, the undiscovered planet beyond Pluto." We are fortunate to live at a time when the "undiscovered" is becoming discovered. Large objects beyond Pluto are now being discovered fairly regularly, and this has led astronomers to attempt to define more carefully which objects are planets. By engaging in discussions of what makes a planet a planet and whether or not Pluto belongs in the "planet" category, students reinforce their knowledge of the characteristics of objects in the Solar System

Students who have completed Units 1 and 2 now have extensive experience with the Sun and how it affects Earth. This background can provide them with extra insight into how Solar System objects share the Sun's environment, the system's central figure. Students who go on to do activities in Unit 4 will see the Solar System as a possible example of many such systems that are abundant in the Galaxy and the Universe.

Credit: NASA/JPL-Caltech

UNIT 3
The Solar System

SESSION SUMMARIES (11 Sessions)

3.1 Models and the Pre-unit 3 Questionnaire

In this session, students learn that a model is a useful tool for understanding how something works. Students also discuss what makes a model accurate or inaccurate. After taking the Pre-unit 3 Questionnaire, the class discusses the accuracies and inaccuracies they found in the depicted model. By the end of this unit, students should be able to identify all of the major inaccuracies shown in the Solar System model on the questionnaire. During this session, the key concept added to the classroom concept wall is:

- *Scientists use models to demonstrate ideas, explain observations, and make predictions.*

3.2 Observing the Jupiter System

In this session, students carefully observe Jupiter's moons—just as Galileo did. The class discusses their observations and arrives at the conclusion that Jupiter is at the center of a system of orbiting moons. This discovery not only highlights an important distinction between the terms *planet* and *moon*, but also provides observational evidence to support the heliocentric model Galileo advocated—one where the Sun, not the Earth, is at the center of the Solar System. During this session, the key concepts that will be added to the classroom concept wall are:

- *Telescopes allow astronomers to observe distant objects in space that might otherwise not be seen.*
- *Scientific explanations are based on evidence gathered from observations and investigations.*
- *Planets orbit stars.*
- *Moons orbit planets.*
- *The Solar System is centered around the Sun—the only star in the Solar System.*

3.3 Exploring Diversity in the Solar System

In this session, students work in teams to examine a set of 36 Solar System object cards. After freely looking at the images, teams divide the set among themselves, with team members sorting their cards into various categories. Afterward, students challenge their team members to discover the criteria they used to sort the images. The main goal of this first encounter with the cards is to excite student curiosity and appreciation for the great diversity of objects in the Solar System. By challenging each other with different ways of sorting a set of cards, student teams learn what defines a category and how objects are placed into categories. Students continue to learn about various Solar System objects by revisiting the cards again in later sessions. During this session, the key concept that will be added to the classroom concept wall is:

- *Many diverse objects make up the Solar System.*

3.4 Categorizing Objects in the Solar System

In this session, groups of students work with the Solar System card sets to identify objects they think might be planets. Teams share the criteria they used to decide whether an object is a planet or not, and the class generates a list of potential planetary characteristics. The class then modifies the list through a guided, whole-class examination of the cards, which reveals additional information about some of the objects. After working through categorizations of various Solar System objects (planets, moons, asteroids, comets, and Kuiper Belt Objects), the class debates the question astronomers themselves are still debating: "Is Pluto a planet?" During this session, the key concepts that will be added to the classroom concept wall are:

- *Scientists categorize objects in the Solar System by characteristics such as: shape and appearance, what they orbit, how large they are, and how far away their orbits are from the Sun.*
- *Not every Solar System object can be easily categorized.*

3.5 Researching Objects in the Solar System

In this session, students learn more about an assigned Solar System object by conducting their own research. Working in pairs, students create interesting and creative learning stations where they can share the results of their research with one another. Each learning station includes a color illustration and scale model of the assigned object, as well as a set of creative travel brochures sharing fun and accurate information about the object. Students prepare these projects for Session 3.7, when the class will tour the Solar System by visiting all of the learning stations to find out more about the different characteristics of various Solar System objects.

3.6 Completing Solar System Travel Brochures

In this session, students continue work on their travel brochures—completing them in preparation for the next session's class tour of the learning stations. Some students may need additional time to complete their brochures—homework and extra class time are both options for extending this session.

3.7 Taking a Tour of the Solar System

In this session, students tour the Solar System by visiting the learning stations made by their classmates. As students read one another's travel brochures, they learn interesting facts about the different objects. After completing their tour, students fill out object characteristic charts, then refer to these charts in a class discussion about general planetary characteristic trends. During this session, the key concepts that will be added to the classroom concept wall are:
- *Solar System objects have a wide variety of characteristics.*
- *In general, the farther away a planet is from the Sun, the colder its temperature.*
- *The inner planets are smaller in size than the outer planets.*
- *The composition of the inner planets is mostly rocky, while the composition of the outer planets is mostly gaseous.*
- *With the exception of Mercury, all the planets have atmospheres.*

3.8 Outdoor Scale Model of the Inner Solar System

In this session, students begin by organizing the planet scale models collected from their learning stations. As they order each planet according to its distance from the Sun, they observe that the planets' orbits naturally divide the Solar System into two regions. The inner Solar System consists of small, rocky planets, while the outer Solar System consists of large, gaseous planets. The class then heads outside to construct an outdoor scale model of the inner Solar System. As students pace out and visualize the orbits of the four inner planets, they begin to realize that the Solar System consists primarily of the empty space between the planets. The class considers the orbits of the outer planets and completes the Solar System scale model in Session 3.9.

3.9 The Outer Solar System and Beyond

In this session, students finish the scale model by looking at the orbits of the outer planets drawn to scale on a map of the school. After discussing the large distances between the outer planets, students contemplate the question, "Where does the Solar System end?" by reading about the NASA Interstellar Boundary Explorer (IBEX) mission. Afterward, student groups discuss how an organized system such as the Solar System might have formed. Using a simple model, students observe how a disorderly collection of matter might have evolved into a more orderly system of orbiting objects. During this session, the key concepts that will be added to the classroom concept wall are:
- *The Solar System is mostly empty space and is very large compared to the objects within it.*
- *The Solar System formed from gas and dust, which gravity drew together into a whirling system.*

3.10 Human-powered Orrery

In this session, the class works together to create a human-powered orrery to model the movements of the four inner planets. Students assist in setting up this moving model of the Solar System and take turns playing the roles of Mercury, Venus, Earth, and Mars. As the class observes the orrery in motion, they form conclusions about the orbital periods of the inner planets. Afterward, the class predicts the orbital periods of the outer planets using the mapped scale model transparency from Session 3.9. During this session, the key concepts that will be added to the classroom concept wall are:
- *Planets closer to the Sun have smaller orbits and move more quickly than planets farther from the Sun.*
- *Objects in the Solar System are in regular and predictable motion.*
- *As seen from Earth, the positions of the planets and the Sun are always changing.*

3.11 Evidence Circles and the Post-unit 3 Questionnaire

In this session, students work in evidence circles to discuss the question of whether or not Pluto is a planet. Although this is a real question circulating among astronomers, the focus of the evidence circles is not primarily on Pluto. It is more a question about whether or not and/or how the category and definition of the term *planet* is useful in organizing objects in the Solar System. This activity gives students the opportunity to synthesize much of the information and concepts that are a part of this unit. The session closes with students filling out the Post-unit 3 Questionnaire. Both the questionnaire and students' written work about categorizing Pluto can be used to assess their learning in the unit.

SESSION 3.1
Models and the Pre-unit 3 Questionnaire

Overview

The questionnaire that launches this unit shows a model of the Solar System. It is extremely difficult to create a completely accurate model, and the one on the questionnaire is no exception. Although it shows correctly that the Sun is in the center of a system of orbiting bodies, it misrepresents the Solar System in many ways. The relative sizes of the objects and the spacing of their orbits are inaccurate. The diagram also fails to include other Solar System objects such as moons and comets. In this session, students learn that a model is a useful tool for understanding how something works. Students also discuss what makes a model accurate or inaccurate. After taking the Pre-unit 3 Questionnaire, the class discusses the accuracies and inaccuracies they found in the depicted model. By the end of this unit, students should be able to identify all the major inaccuracies shown in the Solar System model on the questionnaire. During this session, the key concept that will be added to the classroom concept wall is:

- *Scientists use models to demonstrate ideas, explain observations, and make predictions.*

Models and the Pre-unit 3 Questionnaire	Estimated Time
Introducing Models	5 minutes
Taking the Pre-unit 3 Questionnaire	20 minutes
Discussing the Questionnaire	20 minutes
Total	**45 minutes**

What You Need

For the class:
- ❏ overhead projector or computer with large-screen monitor or LCD projector
- ❏ prepared key concept sheet from the copymaster packet or CD-ROM file
- ❏ two sentence strips
- ❏ a marker
- ❏ transparencies of the three pages of the Pre-unit 3 Questionnaire from the transparency packet or the CD-ROM file

For each team of 4–6 students:
- ❏ 1 copy of the Pre-unit 3 Questionnaire (three pages) from the copymaster packet or the CD-ROM file
- ❏ scratch paper

For each student:
- ❏ 1 copy of the Pre-unit 3 Questionnaire (three pages) from the copymaster packet or the CD-ROM file

Unit Goals

The Solar System is centered around the Sun, the only star in the Solar System.

A wide variety of objects orbit the Sun in the Solar System.

Scientists categorize Solar System objects according to their characteristics; however, not all objects can be easily categorized.

Objects in the Solar System are in regular and predictable motion.

The Solar System is mostly empty space, and is very large compared to the objects located within it.

TEACHER CONSIDERATIONS

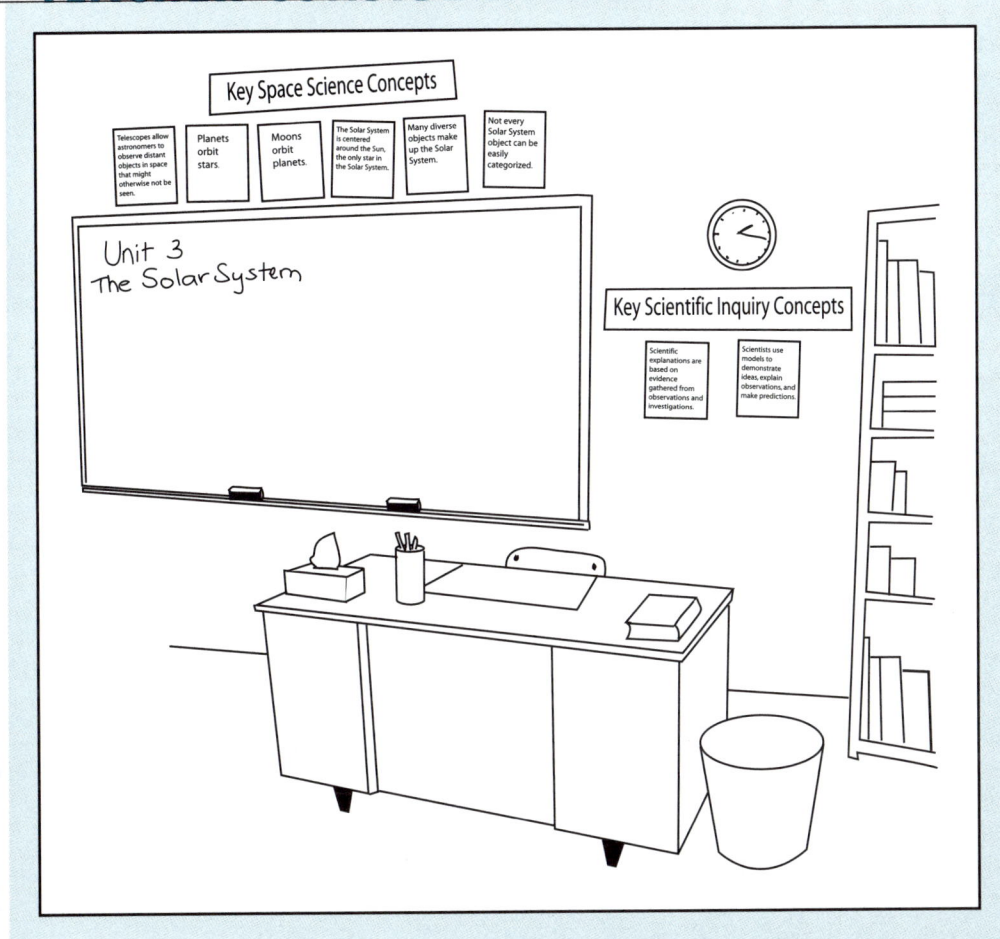

Models are extremely useful in many fields of study—especially in astronomy—when the objects of interest are usually too large, too massive, or too distant to be brought into a laboratory for hands-on study. Students work with several different models throughout this unit.

Key Vocabulary

Scientific Inquiry Vocabulary

Category
Characteristic
Evidence
Model
Observation
Prediction
Scale
Scale model
Scientific explanation

Space Science Vocabulary

Asteroid
Astronomical Unit (AU)
Comet
Diameter
Heliosphere
Kuiper Belt Object (KBO)
Moon
Orbit
Planet
Sphere
Star
System

SESSION 3.1 Models and the Pre-unit 3 Questionnaire

Getting Ready

1. **Designate a wall or bulletin board in the classroom as the unit's concept wall.** Use the concept wall as a space to post the key concepts for each session in this unit. It is helpful to distinguish between the two categories of key concepts—Space Science and Scientific Inquiry. See page 385 for an example of a concept wall. Although the concept wall does take up a fair amount of space, it is a very useful and important learning tool for your students.

2. **Arrange for the appropriate projector format to display images to the class.** Decide whether you will be using the overheads or the CD-ROM. Set up an overhead projector or a computer with a large-screen monitor or LCD projector.

3. **Prepare the key concept sheet.** Make a copy and have it ready to post onto the classroom concept wall during the session.

4. **Prepare the concept wall headings.** Using the sentence strips and marker, make the two headings for the concept wall—one labeled Key Space Science Concepts and the other labeled Key Scientific Inquiry Concepts.

5. **Decide how you will divide the class into teams of 4–6 students.**

6. **Prepare copies of the Pre-unit 3 Questionnaire for the session.** Make a copy of the questionnaire for each student and for each team of students.

GO! Introducing Models

1. **Tell the class that they will be studying the Solar System during this unit.** Ask students to raise their hands if they have seen a model or picture of the Solar System before. Call on students and have them describe the Solar System models they have seen.

2. **Explain what models are and why they are useful.** Tell students that *models* are important tools for studying things that are very big, very small, or very difficult to bring into a classroom or laboratory. If there are any models in your classroom (such as a world globe or DNA molecule), point them out to your students. Explain that although models are not the real objects themselves, they are often very useful for explaining and understanding the real objects they represent.

This discussion about models is meant to be a very brief and introductory one. (Plan to limit it to 5 minutes only.) You may choose to allot additional class time for a more detailed discussion if your students have never worked with models before.

TEACHER CONSIDERATIONS

TEACHING NOTES

The key concepts can be posted in many different ways. If you don't want to use sentence sheets, here are some alternatives:

- Write the key concepts out on sentence strips.
- Write the key concepts out before class on a posted piece of butcher paper. Cover each concept with a strip of butcher paper and reveal each one as it is brought up in the class discussion.

Introducing the Concept of Models. If your students have not been introduced to models before, consider spending some additional time explaining models as outlined below:

Introduce models. If you have a model in your classroom (such as a world globe), hold it up and say that a model is something that represents the real thing. Explain that a model can be used to explain and understand the real object it represents. Tell the class that scientists often use models to better understand something they are studying.

Finding familiar models. Ask students if they have ever played with a toy car or doll house. Explain that these two things are examples of models. Ask students if they can think of any other models they have seen or used before.

Models are not exactly like the real things they represent. Emphasize that while a good model is like the real thing, no model is exactly the same as the real thing. Ask, "What are some ways in which a toy car is not exactly like a real car?" [It's smaller, it has no running motor, the doors don't open, there are no lights, it can't really be driven, etc.] Note: Students enjoy finding inaccuracies in models, and they are usually very good at it!

Define scale models. Tell the class that often a model is "made to scale." What this means is that every part of the real object is measured and made smaller (or larger) by the same factor to create the model. Tell the class that a toy car is often an example of a scale model. Every part of the car has been made the appropriate size relative to other parts of the car. Ask the class to imagine a toy car with a steering wheel larger than the car's tires—it wouldn't be a very good scale model of a real car!

If your students have gone through Unit 2: Why Are There Seasons?, they may remember using a small polystyrene ball as a model of the Earth and a light bulb as a model of the Sun (in Session 2.4) to study seasonal variations in day lengths. While the light bulb worked as a model of the Sun because it cast light on the Earth, it was smaller than the ball that represented the Earth. The Earth and Sun models were useful for studying one aspect of the Earth-Sun relationship, but were not accurate representations of the sizes of the Earth and Sun relative to one another.

SESSION 3.1   Models and the Pre-unit 3 Questionnaire

Pre-unit 3 Questionnaire: The Solar System

1. What are at least two **accurate** things about this diagram as a model of the Solar System?

2. What are three or more **inaccurate** things about this diagram as a model of the Solar System? In the chart below, list and explain how you would fix these inaccuracies.

Inaccurate things about the model:	How you would fix these inaccuracies:

Student Sheet and Transparency—Space Science Sequence 3.1

A scale model is a representation of an object that is smaller or larger than the actual size of the real object. Every part of the real object is measured and made smaller or larger by the same factor to create the scale model.

3. **All models are inaccurate in some way.** Define *accurate* as being correct, with no mistakes or errors. Tell students that since a model is not an identical copy of the real thing, it will have some inaccuracies. Even a good model will not be perfectly accurate. A model that is inaccurate can still be useful because it can accurately represent something else about the real thing. While a world globe is significantly smaller than the actual size of Earth, it is useful for studying the geographical locations of countries relative to one another.

4. **Mention that pictures are models too.** Point out that not all models are three-dimensional. Pictures, illustrations, and diagrams are also models. Although they are flat and two-dimensional, they can still be used to accurately represent certain aspects of an object or thing.

5. **Introduce the concept wall.** Point out the space in the classroom you've designated for the concept wall. Explain that as the class learns important or key concepts about the Solar System, the concept wall will help them to keep track of what they've learned.

6. **Working with models in this unit.** Tell students that throughout this unit they will be working with several different models (some of them scale models) to help them understand and learn more about the Solar System in which we live. Post on the concept wall, under Key Scientific Inquiry Concepts:

   Scientists use models to demonstrate ideas, explain observations, and make predictions.

Taking the Pre-unit 3 Questionnaire

1. **Explain to the students that the questionnaire is not a test.** Students should not feel nervous about doing well and should instead consider the questionnaire as an opportunity to share their ideas about the Solar System. At the end of the unit, they will take the questionnaire again to see if any of their ideas have changed.

2. **Each student should work independently.** Tell students that they should work silently and without helping others. They will have a chance later on to discuss the questions and share ideas with each other.

3. **Display the questionnaire transparency on an overhead projector or a computer monitor.** Tell students that before they begin, you are going to go over a few of the questions in order to clarify them.

TEACHER CONSIDERATIONS

TEACHING NOTES

Some students may feel very anxious about answering questions to which they don't know the answer. Emphasize to your students that the questionnaire is meant to get them to start thinking about the unit's topics, and that it's OK not to know the answers to some of the questions.

Be sure to collect the questionnaires! The Post-unit 3 Questionnaire has the exact same questions as the Pre-unit 3 Questionnaire, with the questions listed in a different order. This is to encourage students to answer thoughtfully when taking the post-questionnaire, rather than to simply remember the order of the questions and their answers from the pre-questionnaire.

QUESTIONNAIRE CONNECTION

Students may point out some of the following things that the model depicts accurately:

- Nine large objects are shown orbiting the Sun. These correspond accurately (in only some ways) to the eight planets and Pluto.
- The Sun is located in the center of the system of orbiting bodies.
- The orbit of the ninth large object (Pluto) crosses the orbit of the eighth planet (Neptune).
- There is an asteroid belt shown between the orbits of the fourth planet (Mars) and the fifth planet (Jupiter).
- The fifth planet (Jupiter) is the largest and has a "Great Red Spot" on it.
- The sixth planet (Saturn) is ringed. (Saturn has a system of many rings. Through a small telescope, they may appear to be a single broad ring, not a single narrow ring as shown in the model.)
- The orbit shape of the ninth body (Pluto) orbiting the Sun is slightly more elliptical than the other orbits.
- The orbits of most of the orbiting bodies are depicted as nearly circular. (This is the case for all of the objects except Mercury, Pluto, and the Asteroid Belt.)

Some examples of inaccuracies students may point out about the Solar System diagram model:

- No moons are pictured around the planets.
- The Sun has "spiky things" around it.
- Distances of planets from the Sun are incorrect.
- Planet sizes are not correct relative to each other or to the Sun.
- Jupiter's "Great Red Spot" is shown on the equator of the planet. (The "Great Red Spot" is actually located in the Southern Hemisphere of the planet.)
- There are no rings shown around Jupiter, Uranus, or Neptune.

continued on page 391

One teacher said, "This introductory lesson piqued the students' interest. I was surprised that only one student commented that the Sun did not really look like it had spikes on it."

SESSION 3.1 Models and the Pre-unit 3 Questionnaire

One teacher said, "The lesson engaged the students in the critical thinking process. The questionnaire was a great attention getter and focus item. In addition, it was very helpful to students when they pulled back together for a whole group discussion. Everyone was able to contribute something."

4. **The first two questions refer to a model of the Solar System.** In Question #1, students should list some things that are *accurate* about the drawing of the Solar System shown. In Question #2, students should list some things that are *inaccurate* about the drawing, as well as how they might correct these inaccuracies

5. **Go over question #3 if you think your students will need guidance on how to fill out the charts.** Explain to students that they will need to check a box for each of the planetary objects listed in the four comparison charts for Question #3.

6. **Pass out the questionnaires.** Give the class 15–20 minutes to complete the questionnaire. Remind students to check that their names are on their papers and then collect the questionnaires.

Discussing the Questionnaire

1. **Arrange the class into teams of 4–6 students.**

2. **Pass out one blank copy of the questionnaire to each team.** Have students arrange themselves around the questionnaire so everyone can see the diagram of the Solar System.

3. **Teams discuss the Solar System diagram.** Each team member should come up with one thing that the model shows *accurately* and one thing that the model shows *inaccurately*. Teams should designate a recorder to list the group's ideas on a piece of scratch paper.

4. **The whole class discusses what is accurate about the diagram (Question #1).** Bring the whole class together and ask teams to share what they found to be *accurate* about the model. Ask, "What are some things about the Solar System model that are accurate?" [The Sun is in the middle of the Solar System, Jupiter has a "Great Red Spot," Saturn has rings, there is an asteroid belt between Mars and Jupiter, etc.]

5. **Call on students to list some inaccurate things about the diagram (Question #2).** Ask students to share what they found to be *inaccurate* about the model. Ask, "What are some things about the Solar System diagram that are inaccurate?" [The orbits of the planets are not spaced apart correctly, the Sun doesn't really have spiky things on it, Jupiter and Saturn almost look as big as the Sun, etc.]

6. **Ask students how they might correct the model's inaccuracies to create a better model of the Solar System (Question #2).** Ask a few students to share their ideas for making the diagram more accurate. Accept all ideas. Tell students that by the end of this unit, they will know much more about the Solar System and how to make a better model.

TEACHER CONSIDERATIONS

QUESTIONNAIRE CONNECTION
continued from page 389

These are some additional inaccuracies about the diagram as a model of the Solar System, most of which students should grasp by the end of the unit:

- The Sun has points like a "storybook Sun."
- The Sun is too small in relation to the planets.
- The Sun is too large in relation to the distances between objects.
- The Sun makes no day or night on the planets.
- All the planets are too large for the sizes of their orbits.
- The four inner planets should be in orbits spaced much closer to each other compared with the orbits of the outer planets.
- The second and third planets (Venus and Earth) should be about the same size, and they should be larger than the first and fourth planets (Mercury and Mars).
- Mercury and Mars have orbits that are slightly more elliptical than the ones shown. (The other planets have orbits that are very nearly circular, as shown.)
- The Asteroid Belt looks like a dotted line. Individual asteroids are much too small to be seen at the scale of the size of the belt in the model.
- Jupiter and Saturn (planets five and six) should be much bigger than the four inner planets and Pluto.
- Jupiter and Saturn should be much smaller than the Sun.
- Saturn has many rings, and other outer planets have rings (although they do not show up easily).
- Other objects as large as Pluto exist beyond Neptune.
- Most planets have moons, some of them larger than Pluto.
- Smaller objects are not on the model, such as moons, comets, asteroids outside the Asteroid Belt, and KBOs (Kuiper Belt Objects).
- The whole model is at room temperature instead of having a super-hot Sun and extremely cold outer planets.
- The orbits of the planets (and Pluto) are visible as rings in space.
- The model is two-dimensional, black and white, and does not move.

One teacher said, "We had a bit of competitive spirit going on between groups for how many accuracies and inaccuracies they could identify on the model. They seemed to enjoy it and came up with a lot of good information."

SESSION 3.2
Observing the Jupiter System

Overview

We cannot observe our Solar System as someone located outside of it might. However, we *can* observe the system of Jupiter and its moons, which can serve as a useful analogy for understanding our own Solar System. In 1610, Galileo's discovery and careful observations of four of Jupiter's moons were instrumental in disproving the geocentric model, which held that the Earth was at the center of everything, including the Solar System. In this session, students carefully observe Jupiter's moons just as Galileo did. The class discusses their observations and arrives at the conclusion that Jupiter is at the center of a system of orbiting moons. Students' reenactment of this discovery highlights the important role of the telescope in the development of astronomy, starting at the very onset of the telescope's use as a scientific instrument. It also clarifies the distinction between the terms *planet* and *moon*. During this session, the key concepts that will be added to the classroom concept wall are:

- *Telescopes allow astronomers to observe distant objects in space that might otherwise not be seen.*
- *Scientific explanations are based on evidence gathered from observations and investigations.*
- *Planets orbit stars.*
- *Moons orbit planets.*
- *The Solar System is centered around the Sun—the only star in the Solar System.*

Observing the Jupiter System	Estimated Time
Introducing Galileo and the Telescope	5 minutes
What Galileo Saw: Observing Jupiter	15 minutes
Reviewing and Discussing the Data	15 minutes
Evidence for the Heliocentric Solar System	10 minutes
Total	**45 minutes**

What You Need

For the class:
- ❑ overhead projector or computer with large-screen monitor or LCD projector
- ❑ prepared key concept sheets from the copymaster packet or CD-ROM file
- ❑ transparency of the Telescopic View of Jupiter With Galileo's Observation Notes from the transparency packet or CD-ROM file
- ❑ transparency of the Observing Jupiter Over 9 Nights from the transparency packet or CD-ROM file

Unit Goals

The Solar System is centered around the Sun, the only star in the Solar System.

A wide variety of objects orbit the Sun in the Solar System.

Scientists categorize Solar System objects according to their characteristics; however, not all objects can be easily categorized.

Objects in the Solar System are in regular and predictable motion.

The Solar System is mostly empty space, and is very large compared to the objects located within it.

TEACHER CONSIDERATIONS

Key Vocabulary

Scientific Inquiry Vocabulary

Category
Characteristic
Evidence
Model
Observation
Prediction
Scale
Scale model
Scientific explanation

Space Science Vocabulary

Asteroid
Astronomical Unit (AU)
Comet
Diameter
Heliosphere
Kuiper Belt Object (KBO)
Moon
Orbit
Planet
Sphere
Star
System

SESSION 3.2 Observing the Jupiter System

- ❏ 4 transparencies of the Tracking Objects Near Jupiter from the transparency packet or CD-ROM file
- ❏ transparency pens in the following colors: black, blue, orange, and red
- ❏ transparencies of the three pages of the Pre-unit 3 Questionnaire from the transparency packet or CD-ROM file

For each student:
- ❏ 1 copy of the Tracking Objects Near Jupiter student sheet from the copymaster packet or CD-ROM file
- ❏ 1 pencil

Getting Ready

1. **Arrange for the appropriate projector format to display images to the class.** Decide whether you will be using the overheads or the CD-ROM. Set up an overhead projector or a computer with a large-screen monitor or LCD projector.

2. **Prepare the key concept sheets.** Make a copy of each key concept and have them ready to post onto the classroom concept wall during the session.

3. **Decide how you will divide the class into teams for the Observing Jupiter activity.** You will need at least four teams—one for each of the moons to be observed.

4. **Make copies of the Tracking Objects Near Jupiter student sheet.** Each student should receive a copy, and you will need to make three transparencies (for a total of four) for the class discussion.

TEACHER CONSIDERATIONS

TEACHING NOTES

The key concepts can be posted in many different ways. If you don't want to use sentence sheets, here are some alternatives:

- Write the key concepts out on sentence strips.
- Write the key concepts out before class on a posted piece of butcher paper. Cover each concept with a strip of butcher paper and reveal each one as it is brought up in the class discussion.

If you plan to use the CD-ROM for this session, be prepared to switch between the CD-ROM and overhead transparencies. If you'd prefer not to have to switch constantly between the two, use the transparencies for the lesson and show the Jupiter's Moons Animation in Galileo's Study (within the Observing Jupiter activity) only if your students need it.

SESSION 3.2 Observing the Jupiter System

▶ Introducing Galileo and the Telescope

1. **The spyglass was invented to look at distant objects.** Tell students that in 1608, makers of eyeglasses in the Netherlands were putting together devices called a spyglass, which consisted of a tube with a lens (a piece of glass with curved sides) at each end. The spyglass could be used to look at distant objects, making them appear larger and closer.

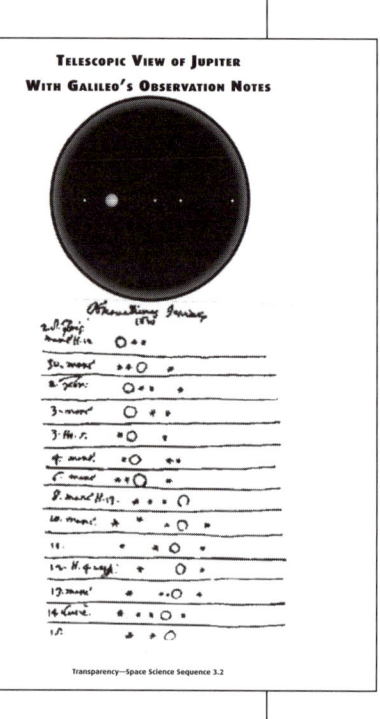

2. **Introduce Galileo and the invention of the telescope.** By 1609, news of the spyglass had reached Galileo Galilei, a scientist in Italy. Galileo was interested in using the spyglass to study objects in the sky. He began creating his own spyglasses, improving upon the design and calling them telescopes.

3. **Have students briefly share their experiences with telescopes.** Ask, "Have any of you looked through a telescope?" Follow up student responses with questions such as, "What did you look at?" "What was it like?" and "Did you look at things in the sky or on the Earth?"

4. **Show the class the Telescopic View of Jupiter with Galileo's Observation Notes overhead transparency. If you are using the CD-ROM, open the Observing Jupiter activity and click on Galileo's Notes.** Tell students that when Galileo observed Jupiter with his telescope, he was the first person to see that a planet is spherical. He also noticed four "specks" of light close to Jupiter. When he first saw them, he wrote in his notebook about how the "stars" near Jupiter were in a neat little line.

5. **Don't reveal to your students yet that these "stars" are actually Jupiter's moons!** At some point, your students will probably realize that the four spots near Jupiter are actually its orbiting moons. If they bring this up at any time during the session, confirm that they are indeed the moons of Jupiter, and that Galileo eventually came to the same conclusion. From then on, refer to the spots as moons. However, if no one points this out, continue to refer to the spots as "specks" or "stars" and wait to reveal that they are Jupiter's moons until later in this session (during Reviewing and Discussing the Data).

6. **Explain that Galileo observed and recorded the movements of these spots carefully.** Tell students that Galileo watched these "star-like" objects and noted that as Jupiter moved across the sky, the "star-like" objects moved with it. He also observed that they appeared in different places around Jupiter each night. Explain that this was very unusual, since other stars never changed position in relation to one another.

396 • SPACE SCIENCE SEQUENCE 6–8   Session 3.2: Observing the Jupiter System

TEACHER CONSIDERATIONS

TEACHING NOTES

Although Galileo is sometimes credited as the inventor of the telescope, he was not its inventor. However, it is not entirely clear who actually did invent it. Hans Lippershey, a Dutch spectacle maker, applied for the first patent in 1608, but two others—Jacob Metier and Zacharias Janssen, who were also Dutch spectacle makers—applied a short time later. They all claimed to have invented it first and accused the others of stealing the idea. The idea had actually been written about hundreds of years earlier, but the technology of making glass and lenses was not far enough advanced at that time for any device to be made. Galileo does, however, deserve credit as the first person to make major discoveries using a telescope to study the night sky. These include the discovery of the moons of Jupiter, the surface features on the Moon, sunspots, the phases of Venus, and stars in the Milky Way.

One teacher said, "I emphasized the importance of keeping records in their science notebooks. We had a big laugh over the fact that Galileo probably had a science teacher who taught him to write everything down. It was a very engaging lesson for all of us."

SESSION 3.2 Observing the Jupiter System

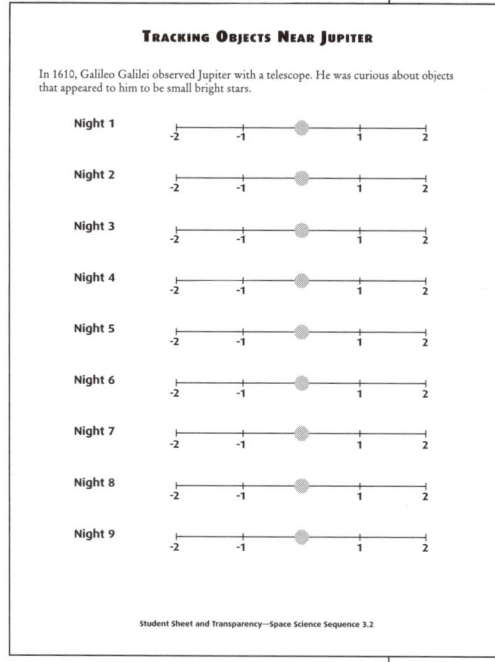

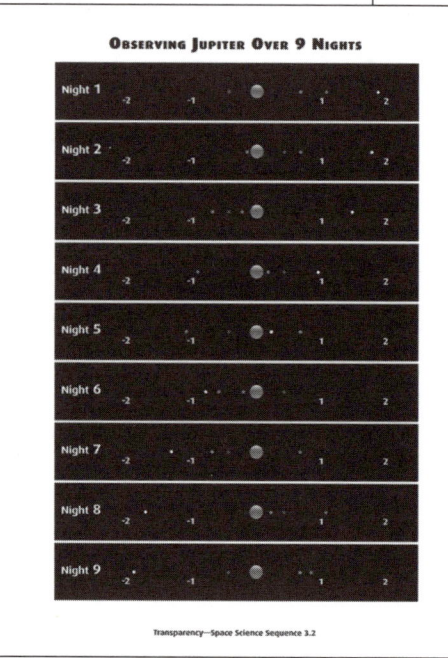

7. Explain to students that Galileo's telescope allowed him to observe a distant object like Jupiter. Tell students that without a telescope, Galileo would not have been able to observe Jupiter in much detail. Today, they will observe Jupiter just as Galileo did and learn what he discovered from his careful observations. Post on the concept wall, under Key Space Science Concepts:

Telescopes allow astronomers to observe distant objects in space that might otherwise not be seen.

What Galileo Saw: Observing Jupiter

1. Pass out the Tracking Objects Near Jupiter student sheets and pencils. Each student should receive an observation sheet and pencil. Tell students that, on this sheet, they will record their observations of Jupiter as though they had been observing it over nine consecutive "nights."

2. Begin by showing the class Night 1 only. Put the Observing Jupiter Over 9 Nights transparency (with the moons already colored) on the overhead projector. Show only Night 1, **covering all other nights with a piece of paper**. If you are using the CD-ROM, open the Observing Jupiter activity and click on Look Through the Telescope to reveal Night 1.

3. Explain why the spots around Jupiter have been colored in. Explain to the class that what they are seeing represents a view of Jupiter through a telescope with a little more information than what Galileo had. The four "star-like" objects have been given different colors so they can be more easily distinguished from one another. Night 1 indicates that this is the first night of Galileo's observations.

TEACHER CONSIDERATIONS

CD-ROM NOTES

CD-ROM Instructions: Galileo's Study

Three activities can be found in Galileo's Study, which is within the Observing Jupiter activity on the CD-ROM:

1. **An animation of Galileo's observations of Jupiter's moons.** (You may choose to use this in place of the overhead transparency for the Observing Jupiter activity.) Click on the telescope to access Galileo's observations. Use the BACK and NEXT arrows to navigate through each of the nine nights. Click on MAIN MENU to return to the study.

2. **An animated orrery of Jupiter and its moons.** Click on the Jupiter system model sitting on the table to access this interactive. The animation shows synchronized views of the movements of Jupiter's moons from two perspectives—the top and the side. (Elapsed time is counted in Earth days.) Click PLAY to begin the animation. This is an excellent way to demonstrate orbital movement to your students if they are having difficulty visualizing the orbits of Jupiter's moons from different perspectives. Click on MAIN MENU to return to the study.

3. **A copy of Galileo's observation notes.** Click on the book on Galileo's table to access his actual, handwritten observations of the moons of Jupiter. To enlarge the interactive to full screen, press CONTROL F (Windows) or APPLE F (Macs). Press ESC to exit the full-screen display. Click on MAIN MENU to return to the study.

SESSION 3.2 Observing the Jupiter System

4. **Explain the number line.** Point out that the numbers to the right and left of Jupiter are like the numbers on a number line with Jupiter in the position of zero. Students should use these numbers to describe the positions of the colored spots. To assess your students' understanding of this, ask them the following questions:

   - "What number is the white spot closest to?" [2.]
   - "Which two colored spots are close to the number 1?" [The orange and blue spots.]
   - "Which spot is on the side with the negative numbers?" [The red spot.]
   - "What is located in the place where zero would be on the number line?" [Jupiter.]

5. **Divide the class into teams and assign a colored spot for each team to observe.** Ask each team to imagine that they are a group of astronomers working together to carefully observe their assigned spot. Have them fill in the color of their assigned spot as indicated on their Tracking Objects Near Jupiter student sheet.

6. **Students record their observations for Night 1.** Have students find the positions of their assigned spots for Night 1. Ask them to make an X with their pencils on the Night 1 line to show the position of their spot as they see it in relation to Jupiter and the numbers on the number line. Encourage students within each team to help one another. Before going on to Night 2, walk around the classroom and check to see that everyone understands how to correctly record the position of their assigned spot.

7. **Ask, "Where is your spot on Night 2?"** Cover up Night 1 on the transparency with another piece of paper and move the first piece of paper down to reveal Night 2. (If you are using the CD-ROM, click on the NEXT arrow to move on to Night 2.) Have students record the position of their spots for Night 2, making sure that they mark their observations on the Night 2 line of the student sheet.

8. **Continue moving along from night to night.** Continue revealing each successive nightly view of Jupiter. As you show each night's view, have students record the new position of their spot *on the next lower line*. (Check to make sure that no one is marking all of their observations on the same line.) By Night 4, most students should be proficient at observing and recording their spots, and you can probably pick up the pace.

Optional: For younger students, you might want to turn the classroom lights on and off between observation nights to represent a passing day.

400 • SPACE SCIENCE SEQUENCE 6–8   Session 3.2: Observing the Jupiter System

TEACHER CONSIDERATIONS

TEACHING NOTES
Classroom Management Considerations

- **Have each team of students observe just one of the moons.** This makes it easier for teammates to check one another's observations and to discuss their predictions together.
- **Make sure each moon has a team assigned to observe it.** Designate at least four teams so that observations can be made for each of the moons.
- **The white and orange moons may be difficult to distinguish from one another.** You might want to assign the white and orange moons—Callisto and Europa, respectively—to teams seated near the front of the classroom, since students farther back may have difficulty distinguishing between these two lighter colors.
- **Some students may be able to make observations for more than one moon at a time.** Consider suggesting this as an extra challenge to selected or advanced students, but tell them to give priority to their assigned moon when making their observations.
- **An advanced option for the Observing Jupiter activity.** This is a good option for assigning moon observations if you have older students who work well independently and have good recording skills. Divide the class into groups of four. Ask each member of a group to observe and record a different colored spot. As students progress through each night of observation, they can see how their recorded data differ from those of their team members.

SESSION 3.2 Observing the Jupiter System

9. **Have students predict where they think their spots will be next.** Starting around Night 4 or 5, ask your students to predict where they think their spots will be on the next night. Some students may be doing this already.

10. **Have students compare data with their teammates.** Students should discuss any differences in their observations with teammates as they continue through the activity.

11. **After Night 9, have students connect the Xs on their observation sheets.** Each student should draw a line from the X they made on Night 1 to the X they made on Night 2, continuing down through Night 9. (Each of the four moons will generate a different zigzag pattern as students connect their Xs.)

12. **Record observation results for each spot on transparencies.** Ask four students who observed different colored spots to trace their zigzag patterns onto four separate transparencies of the Tracking Objects Near Jupiter student sheet. Give each student a transparency marker that matches the color of his or her spot. Use a black pen for the white moon (Callisto).

Reviewing and Discussing the Data

1. **If necessary, reveal now that the four "spots" are actually Jupiter's moons.** Ask students, "Are these spots really stars?" "What makes you think that?" Accept responses and, if necessary, explain that the motion of the spots is not like the motion of stars. They have been tracking the movements of the four largest moons of Jupiter, called the Galilean Moons in honor of their discoverer, Galileo.

   - The white moon is Callisto (ka-LIS-toe).
   - The blue moon is Ganymede (GAN-ee-meed).
   - The orange moon is Europa (yur-OH-pa).
   - The red moon is Io (EYE-oh).

   A closer look at the motions of these objects will show why they are classified as moons.

2. **Teams discuss their moon's changes in position.** Ask each team to discuss how their moon moved over nine nights of observation. Ask them to try to explain why their moon changed positions. What might explain their observations?

TEACHER CONSIDERATIONS

SESSION 3.2 Observing the Jupiter System

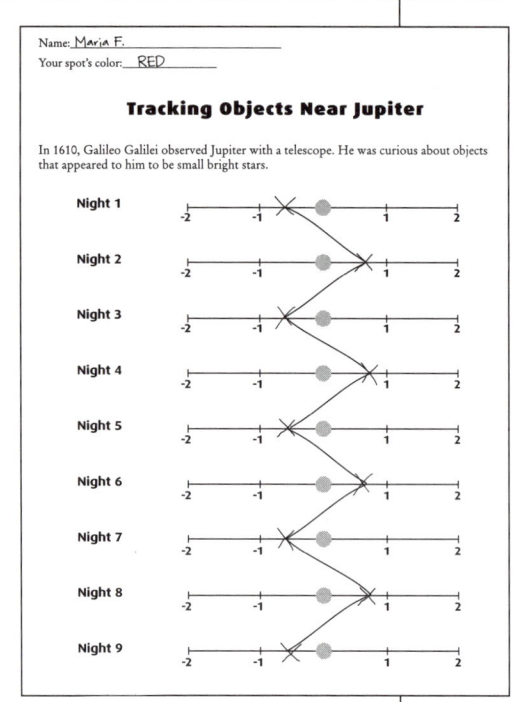

3. **The class discusses Io's movements.** After a few minutes, regain the attention of the class. Display the transparency of the Tracking Objects Near Jupiter student sheet for Io (with its zigzag path marked in red). Ask the team(s) who tracked Io to explain the moon's movements. Ask, "Does Io hop from side to side around Jupiter each night?" [No. Io doesn't really hop around Jupiter each night. It just orbits the planet so quickly that it appears on the other side of Jupiter each night.]

4. **Display the zigzag paths of the other moons one at a time.** For each moon, place the student-prepared transparency tracing the moon's nightly position on the overhead. Ask the team(s) who tracked that moon if they can explain the moon's change in position each night. Ask, "What is causing the moon to change position each night?" [The moon is orbiting Jupiter.]

5. **Solidify student understanding about orbital movement.** Many students may have difficulty deducing their moon's orbital movement from their side-view observations. Invite them to explain how an orbiting object might appear to make the side-to-side motion that they have observed. Possible explanations include:

- The moons are moving around Jupiter like cars on a flat racetrack, and we are viewing the race from the sidelines. If we were to view the race from the elevated box seats, we would see the cars moving around and around the racetrack.
- We are viewing Jupiter and its moons "from the side." If we were looking "from the top," we would see the circular path each moon makes as it moves around Jupiter.

6. **Demonstrate orbital movement using two students.** Remove the transparencies from the overhead projector and have a student stand in the middle of the projected light so his or her shadow falls on the screen. Have another student orbit, or walk around, the first student. The shadow of the orbiting student will move from side to side on the screen even though the student is actually moving in a circle. Ask, "If someone wanted to see the orbiting student going in a circle around the first student, from what location would they have to watch?" [They would have to observe the orbiting student from the ceiling.]

TEACHER CONSIDERATIONS

TEACHING NOTES
Orbital Periods of the Moons. While it's possible for students to use their observation data to calculate the approximate number of nights it takes each moon to orbit Jupiter, their results may vary depending on their observing and recording skills and the method they use to count nights. It's much more important that students understand the orbital movements of the moons themselves, rather than how many nights it takes each moon to orbit Jupiter. Astronomers have determined the following orbital periods for each moon: Io (2 days), Europa (4 days), Ganymede (7 days), and Callisto (17 days).

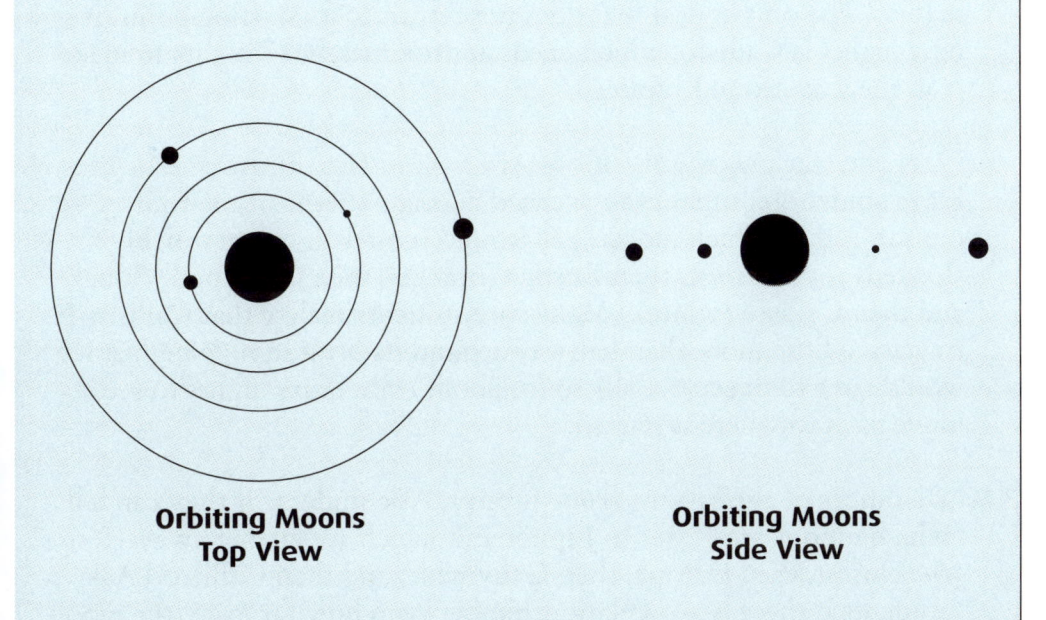

**Orbiting Moons
Top View**

**Orbiting Moons
Side View**

SESSION 3.2 Observing the Jupiter System

Teachers found this animation particularly useful for solidifying student understanding of orbital movement. It's worth showing if you have access to a computer and some additional class time!

7. **If you are using the CD-ROM, click on Jupiter's Moons Animation (in Galileo's Study) within the Observing Jupiter activity to demonstrate orbital movement to your students.** Ask students to observe the movement of Jupiter's moons from both the top and side views.

8. **Note whether moons return to their starting points.** Ask, "Did any of the moons return to their starting points during the nine nights of observation?" [Io returns to its starting point every other night; Europa takes about four nights; Ganymede returns after seven nights.] The only moon that does not return to its starting point over nine nights is Callisto, which needs approximately 17 nights to make it all the way around Jupiter.

9. **Place the stack of all four moon transparencies on the overhead.** Have students compare the overlaid zigzag patterns of the four moons. Ask, "Which moons get around Jupiter the most quickly?" [Io (red) is the fastest, then Europa (orange), then Ganymede (blue), and then Callisto (white).] Make sure students realize that Callisto traveled a little more than halfway around its orbit in nine nights; we would have to observe Callisto for about eight *more* nights to see it move back to where it started.

10. **Distances of the moons from Jupiter.** Ask students if they can tell which moons are closer to Jupiter and which are farther away. [Io is the closest, then Europa, then Ganymede, and then Callisto.] Ask students if there is any relationship between how far away the moon is from Jupiter and how long it takes the moon to orbit Jupiter. Students should notice that the closer the moon is to Jupiter, the faster it completes its orbit around the planet.

TEACHER CONSIDERATIONS

PROVIDING MORE EXPERIENCE
Ask, "If Callisto is farthest from Jupiter, why does it appear to be closer to Jupiter than to Europa on Night 5?" [Certain orbital positions of Europa and Callisto, if viewed from the side, can make it appear that Callisto is closer to Jupiter than to Europa. If Europa has reached the "rightmost" or "leftmost" part of its orbit, AND Callisto happens to be close to crossing in front of or behind Jupiter in its orbit, then it would appear that Callisto is closer to Jupiter than to Europa from a sideview perspective.] It may help to use three students and the overhead to model this for the class. Have one student stand in the middle of the overhead's projected light so that his or her shadow falls on the screen. Ask the second student to walk in a small orbit around the first student. While the second student is orbiting the first student, have the third student walk in a much larger orbit around the first student. Point out when the third student's shadow appears to be closer to the first student's shadow than the second student's shadow.

TEACHING NOTES
Students may wonder whether the moons closer to Jupiter really do orbit faster or whether they simply have less distance to cover because their orbits are smaller. Let students know that they will be exploring this question later on in the unit. In Session 3.10, students will participate in a human orrery and learn that not only do planets closer to the Sun have a smaller orbit to travel, but that those planets actually do travel faster as well.

How Galileo Tracked Jupiter's Moons. Could Galileo see color differences between the four moons? No. For this activity, we've arbitrarily color-coded the moons to make them easier to distinguish from one another, but Galileo would not have had this artificial enhancement to help him with his observations. So how did Galileo track the moons and draw his conclusions? Several things might have helped him. Callisto (the white spot) moves so slowly that it is often very nearly in the same position night after night—making it easy to distinguish this moon from the others. Io (the red spot) is also easily identified, since it moves so quickly in its orbit that a change in its position is detectable after only a few hours of observation. Remember also that Galileo was a seasoned scholar and scientist. Once he realized that the "specks of light" did not behave like stars, he logically concluded that they were most likely the moons of Jupiter. With this conclusion in mind, he could deduce the orbital patterns of each of the four moons.

SESSION 3.2 Observing the Jupiter System

Evidence for the Heliocentric Solar System

1. **Tell students about the geocentric model.** Explain that at one point in time, many people believed that the Earth was located in the center of the Universe with everything else, including the Sun and other planets, revolving around it. Some of your students may have heard of the geocentric model, while others may not be aware of it. If you have additional time, continue discussion of the geocentric model as suggested on the right-hand page under Teacher Considerations.

2. **Discuss how Galileo's observations challenged the geocentric model.** Ask students if they can see how Galileo's observations might have convinced others that the Earth was *not* in the center of the Universe. [Galileo's discovery of moons orbiting Jupiter (and not Earth) suggested that not everything moved around the Earth.]

3. **Galileo used the Jupiter system as a *model* to understand the Solar System.** Explain that Galileo used his observations of the Jupiter system to help him better understand another system—the Solar System. Using the Jupiter system as a model, Galileo reasoned (correctly) that the Solar System was organized similarly to Jupiter and its moons. Remind students of the key concept from Session 3.1:

    Scientists use models to demonstrate ideas, explain observations, and make predictions.

4. **Galileo used his observations to support the heliocentric theory.** Galileo concluded that just as Jupiter's moons orbit Jupiter, the Earth must also orbit something—the Sun. Tell students that the Jupiter system they observed today is like a miniature Solar System, with Jupiter playing the part of the Sun, and Jupiter's moons playing the parts of the planets. Tell the class that Galileo used his observations of Jupiter's moons as evidence for the heliocentric theory. Post on the concept wall, under Key Scientific Inquiry Concepts:

    Scientific explanations are based on evidence gathered from observations and investigations.

TEACHER CONSIDERATIONS

PROVIDING MORE EXPERIENCE
Optional Discussion: Occultation and Transit

1. **Show students the Telescopic View of Jupiter With Galileo's Observation Notes transparency again.** Ask them to notice any similarities or differences between Galileo's notes and their own records.
2. **Tell the class that Jupiter has more than four moons.** Ask the class, "Why do you think Galileo observed the movements of only a few of Jupiter's moons?" "Where were Jupiter's other moons?" [Moons that happened to be in their orbits behind Jupiter could not be seen.]
3. **Define *occultation* and *transit*.** Explain to students that an *occultation* occurs when one object is hidden from view by another object in front of it. A *transit* occurs when an object passes in front of another object.
4. **Moons in occultation or transit with Jupiter would have been difficult to see.** Explain that Galileo would not have been able to see moons behind or in front of Jupiter. Jupiter's brightness would have made it difficult to detect a moon in front of it.

TEACHING NOTES

Additional Classroom Discussion: Examining "Evidence" for the Geocentric Model. Have students consider why so many people in Galileo's time believed that the Earth was in the center of the Universe with the Sun, other stars, and other planets all orbiting around it. Ask, "What evidence could make a person think that?" [It *looks* like everything is moving around the Earth. The Sun and stars seem to revolve in the sky around us. Students might add that even though the Earth is spinning, we can't feel the motion of the Earth. It feels like the Earth is stationary.]

Defining the term *system*. In the most general sense, a *system* is defined as "a group of independent but interrelated elements or objects comprising a unified whole." (The word derives from Greek and Latin, meaning *to combine or place together*.) This definition has been applied in our language in many different contexts. For example, our bodies have digestive and circulatory systems, while science and other disciplines use classification systems. A person can even devise a whole new system for doing a task! Because of the widespread use of this term, you may choose to discuss with your students what they think the term *system* means. If necessary, help them to come up with a useful definition. This will ensure that all students in the class have the same understanding of the term as they discuss the Jupiter system and the Solar System in this and future sessions.

SESSION 3.2 Observing the Jupiter System

5. **The Sun is a star.** Remind students that the Sun is a star. Explain that Earth is just one of a number of planets that orbit around the Sun. Post on the concept wall, under Key Space Science Concepts:

   Planets orbit stars.

6. **Moons orbit planets.** Tell students that just as planets orbit stars, moons orbit planets. Earth's moon orbits Earth, just as Jupiter's moons orbit Jupiter. Post on the concept wall, under Key Space Science Concepts:

   Moons orbit planets.

7. **The Sun is the only star in the Solar System.** Ask the class, "How many stars do you think there are in the Solar System?" [Only one: the Sun.] Tell students that although the Solar System has many systems of planets and moons, they all orbit around the Sun. Post on the concept wall, under Key Space Science Concepts:

   The Solar System is centered around the Sun—the only star in the Solar System.

8. **Revisit Question #4 on the Pre-unit 3 Questionnaire with students.** Show the transparency of the Pre-unit 3 Questionnaire. Go through each true-or-false question and ask students what answers they would choose after today's activity and discussion.

TEACHER CONSIDERATIONS

TEACHING NOTES
Galileo's discovery of the moons of Jupiter as evidence for the heliocentric model of the Solar System is a fascinating chapter in the history of science. For more information on this topic, see the Background Information for Teachers section on page 243.

Defining and classifying objects, such as planets or moons, according to their orbits (and what they orbit around) is a useful way to organize and learn about objects in the Solar System.

QUESTIONNAIRE CONNECTION
Use this opportunity to assess your students' understanding of this session's key concepts. Save the Pre-unit 3 Questionnaire transparency for use with other Questionnaire Connection opportunities in the unit.

SESSION 3.3
Exploring Diversity in the Solar System

Overview

A great variety of objects can be found in the Solar System. In addition to planets and moons, comets and asteroids are also Solar System objects. All of these objects can be characterized according to their size, appearance, composition, and shape. In this session, students work in teams to examine a set of 36 Solar System object cards. After freely exploring the images, teams divide the set among themselves, with team members sorting their cards into various categories. Afterward, students challenge their team members to discover the criteria they used to sort the images. The main goal of this first encounter with the cards is to excite student curiosity and appreciation for the great diversity of objects in the Solar System. By challenging each other with different ways of sorting a set of cards, student teams learn what defines a category and how objects are placed into categories. Students continue to learn about various Solar System objects by revisiting the cards again in later sessions. During this session, the key concept that will be added to the classroom concept wall is:

- *Many diverse objects make up the Solar System.*

Exploring Diversity in the Solar System	Estimated Time
Introduction and Free Exploration of the Solar System Cards	15 minutes
Sorting the Solar System Cards	30 minutes
Total	**45 minutes**

What You Need

For the class:
- ❏ (optional) computer with large-screen monitor or LCD projector
- ❏ (optional) Space Science Sequence CD-ROM
- ❏ prepared key concept sheet from the copymaster packet or CD-ROM file
- ❏ 1 copy of the Key for the Solar System Cards from the copymaster packet or the CD-ROM file
- ❏ 8 different colored markers

For each team of 4-6 students:
- ❏ 1 set of 36 Solar System Cards from the copymaster packet or the CD-ROM file
- ❏ 1 sandwich bag or envelope (to hold a set of 36 cards)

Unit Goals

The Solar System is centered around the Sun, the only star in the Solar System.

A wide variety of objects orbit the Sun in the Solar System.

Scientists categorize Solar System objects according to their characteristics; however, not all objects can be easily categorized.

Objects in the Solar System are in regular and predictable motion.

The Solar System is mostly empty space, and is very large compared to the objects located within it.

TEACHER CONSIDERATIONS

TEACHING NOTES

The key concepts can be posted in many different ways. If you don't want to use sentence sheets, here are some alternatives:

- Write the key concepts out on sentence strips.
- Write the key concepts out before class on a posted piece of butcher paper. Cover each concept with a strip of butcher paper and reveal each one as it is brought up in the class discussion.

Key Vocabulary

Scientific Inquiry Vocabulary

Category
Characteristic
Evidence
Model
Observation
Prediction
Scale
Scale model
Scientific explanation

Space Science Vocabulary

Asteroid
Astronomical Unit (AU)
Comet
Diameter
Heliosphere
Kuiper Belt Object (KBO)
Moon
Orbit
Planet
Sphere
Star
System

SESSION 3.3 Exploring Diversity in the Solar System

Getting Ready

1. **Arrange for the appropriate projector format to display images to the class.** Decide whether you will be using the overheads or the CD-ROM. Set up an overhead projector or a computer with a large-screen monitor or LCD projector.

2. **Prepare the key concept sheet.** Make a copy of the key concept and have it ready to post onto the classroom concept wall during the session.

3. **Decide how you will divide the class into teams of 4-6 students.**

4. **If you have the Space Science Sequence materials kit, find the Solar System card sets (bordered in green) and skip to Step #6.** If you want to color-code the card sets from the kit in order to keep track of each set, see Step #5b before moving on to Step #6.

5. **Prepare the Solar System card sets.** Each set contains 36 cards displaying images of various Solar System objects. The card sets will be introduced during this session and used again in later sessions. For each team of 4-6 students:

   a. **Print out the cards from the CD-ROM file or make color photocopies of the Solar System Cards from the copymaster packet.** You will need one set of 36 cards for each team of students.

   b. **Color-code the card sets.** This will make it easier to keep track of the cards in each set. Using a colored marker, scribble across the back of each set, making sure that you mark the back of every card. Be sure to use a different color marker for each set.

   c. **Cut the cards apart.** Put each set of 36 cards in a sandwich bag or an envelope. If using envelopes, consider labeling them with the same color as the markings on the backs of the cards. With each set of cards, there are four cards containing the names of the space objects. **Do not** put these cards into any of the bags/envelopes for the teams. *These cards are intended for your use only in this session.* You may choose whether you'll have students use them in later sessions.

6. **Familiarize yourself with the images on the cards.** Although you do not need to have expert knowledge about the 36 objects before presenting them to your students, it can be helpful to gain some familiarity with them first. Look over the information on the cards and read the Key for the Solar System Cards on page 417. You may also choose to make a copy of the Key for the Solar System Cards (from the CD-ROM file or copymaster packet) for your reference in this and later sessions. Also read About the Solar System Cards on page 419.

Be sure not to include the cards with the names of the objects in the sets of cards you assemble for the students.

TEACHER CONSIDERATIONS

CD-ROM NOTES
Card-Sort Instructions (for Units 3 and 4).

The Space Science Sequence CD-ROM provides two interactive card-sort activities for use with Units 3 and 4. You may use the card-sort interface as a visual aid for student teams to show the entire class their card sorts or as a discussion tool to lead class-wide sorts.

The Solar System Gallery can be used to show students enlarged versions of objects from the card sets. WARNING: Object names are shown on the cards in the gallery. You might want to show these to your students only after the names of the objects have been revealed.

The Solar System Card-Sort Activity Instructions

1. On the first screen, select the number of category groups needed. Enter in a name for each group and then click on BEGIN to start the sort.

2. On the sorting screen, choose a group number from the left to which to add cards. Objects can be added to a group by clicking on the object image. (Notice that the numbers shown on the images are identical to those on the object cards.) After a group has been chosen, the group number is highlighted. Any cards added to the group take on a border of the group's designated color. To deselect a card from the group, simply click it again. (The group's corresponding colored border will disappear from the object image.) To begin sorting into another group, simply select a new group number from the left.

3. Click on CLEAR to remove any sorts you have created. Click on START OVER to remove your sorts and to go back to the first screen to redesignate and rename groups. (Note: All images do not have to be sorted in order to view your sorts! Once you have completed sorting, click on VIEW SORT to see the images you have sorted into each named group. To modify any of your sorts, click on GO BACK.)

4. To enlarge the interactive to full screen, use CONTROL F for Windows and APPLE F for Macs. Press ESC to exit.

SESSION 3.3 Exploring Diversity in the Solar System

7. **Optional:** If you plan to use the Space Science Sequence CD-ROM, set up the computer with large-screen monitor or LCD projector. The CD-ROM has a useful interactive activity—The Solar System Card Sort—which teams may use to show their sorts to the class. Instructions for using the CD-ROM can be found on page 415.

GO! **Introduction and Free Exploration of the Solar System Cards**

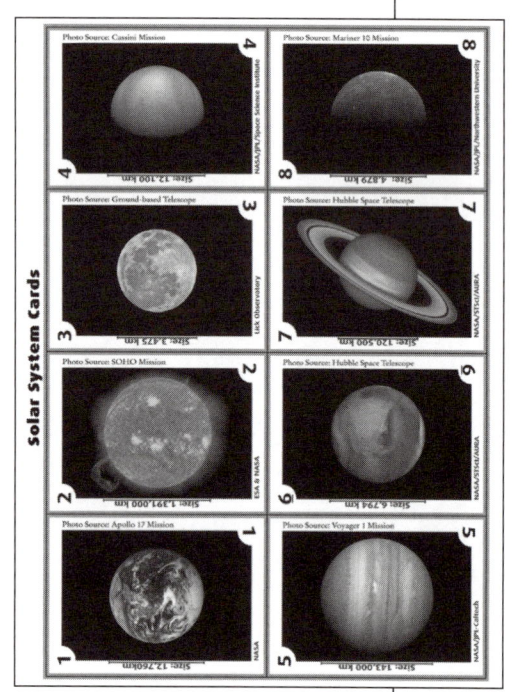

1. **Introduce the card sets.** Ask students what Solar System objects they studied during the last session. [Jupiter and four of its moons—Io, Europa, Ganymede, and Callisto.] Tell students that our Solar System includes many other objects besides Jupiter and its moons. Today, they will have a chance to view some of these other Solar System objects. Explain that just as Galileo was able to observe Jupiter and its moons with his telescope, scientists today continue to actively explore the Solar System using telescopes and space probes. In this session, the class will get to study 36 Solar System object cards displaying images of various objects these scientists have gathered.

2. **Students should handle the cards carefully.** Tell students that because they will be using the same card sets for several sessions, it is very important that they keep each card set complete and in good condition. Ask students to avoid wrinkling or dropping their cards. Also, they should not mix cards from different sets. Point out that the color coding on the backs of the cards should make it easier to distinguish sets from one another.

3. **The cards share limited information about the pictured objects.** Tell students that they will not know the name of the object or even what kind of object it is. Explain that by looking at an object without knowing its name, students will be exploring an image without much additional information, as scientists discovering a new object might do. Tell students that, for now, the cards only tell them:

   - What the object looks like. Each card shows a picture of the object.
   - How big the object is. The size of the object is indicated beside the image.
   - How the object's image was obtained. The card lists the device that took the picture—whether it was a space probe, an Earth-orbiting telescope (such as the Hubble Space Telescope), or a ground-based telescope.

416 • SPACE SCIENCE SEQUENCE 6–8   Session 3.3: Exploring Diversity in the Solar System

TEACHER CONSIDERATIONS

KEY FOR THE SOLAR SYSTEM CARDS

Card Number		Body that it Orbits	Size		Average Radius of Orbit	Year of Discovery
1	Earth	Sun	12,760 km	diameter	149,600,000 km	Ancient times
2	Sun	Galactic Center	1,391,000 km	diameter	250,000,000,000,000 km	Ancient times
3	Moon	Earth	3,475 km	diameter	384,400 km	Ancient times
4	Venus	Sun	12,100 km	diameter	108,200,000 km	Ancient times
5	Jupiter	Sun	143,000 km	diameter	778,400,000 km	Ancient times
6	Mars	Sun	6,794 km	diameter	227,900,000 km	Ancient times
7	Saturn	Sun	120,500 km	diameter	1,427,000,000 km	Ancient times
8	Mercury	Sun	4,879 km	diameter	Elliptical—min: 46,000,000 km max: 69,820,000 km	Ancient times
9	Comet Halley	Sun	16 km	length	Highly Elliptical—min: 87,700,000 km max: 5,250,000,000 km	Ancient times
10	Ganymede	Jupiter	5,268 km	diameter	1,070,000 km	1610
11	Callisto	Jupiter	4,806 km	diameter	1,883,000 km	1610
12	Io	Jupiter	3,650 km	diameter	422,000 km	1610
13	Europa	Jupiter	3,130 km	diameter	671,000 km	1610
14	Titan	Saturn	5,150 km	diameter	1,222,000 km	1655
15	Tethys	Saturn	1,060 km	diameter	294,700 km	1684
16	Uranus	Sun	51,120 km	diameter	2,871,000,000 km	1781
17	Ceres	Sun	950 km	diameter	414,700,000 km	1801
18	Neptune	Sun	49,530 km	diameter	4,498,000,000km	1846
19	Triton	Neptune	2,704 km	diameter	354,800 km	1846
20	Ariel	Uranus	1,160 km	diameter	190,900 km	1851
21	Comet Tempel 1	Sun	7 km	length of nucleus	Highly Elliptical—min: 225,000,000 km max: 708,000,000 km	1867
22	Phobos	Mars	27 km	length	9,378 km	1877
23	Ida	Sun	58 km	length	428,000,000 km	1880
24	Eros	Sun	33 km	length	Elliptical—min: 169,500,000 km max: 266,800,000 km	1898
25	Pheobe	Saturn	220 km	diameter	12,950,000 km	1898
26	Comet Borrelly	Sun	8 km	length of Nucleus	Highly Elliptical—min: 202,000,000 km max: 872,000,000 km	1904
27	Pluto	Sun	2,302 km	diameter	Elliptical—min: 4,437,000,000 km max: 7,376,000,000 km	1930
28	Charon	Pluto	1,186 km	diameter	19,570 km	1978
29	Comet Wild 2	Sun	5 km	diameter	Highly Elliptical—min: 238,000,000 km max: 794,000,000 km	1978
30	Comet Shoemaker-Levy 9		First it orbited the Sun, and then it orbited Jupiter. It broke into several fragments and collided with Jupiter. It no longer exists.			1993
31	Dactyl	Ida	1.5 km	diameter	~90 km	1993
32	Comet Hale-Bopp	Sun	~40 km	Diameter of Nucleus	Highly Elliptical—min: 137,000,000 km max: 55,500,000,000 km	1995
33	Itokawa	Sun	0.5 km	length	Elliptical—min: 142,600,000 km max: 253,500,000 km	1998
34	Quaoar	Sun	~1,300 km	diameter	6,500,000,000 km	2002
35	Eris	Sun	~2,500 km	diameter	Elliptical—min: 5,650,000,000 km max: 14,600,000,000 km	2003
36	Sedna	Sun	~1,500 km	diameter	Highly Elliptical—min: 11,390,000,000 km max: 145,900,000,000 km	2004

The cards do not indicate the names of the objects. Do not reveal object names to the students until directed to do so! The goal of this session is to encourage students to explore the images freely, without knowledge of what the object is.

SESSION 3.3 Exploring Diversity in the Solar System

4. **Cards display the best images currently available.** Let students know that each card image is the best one available of an object as it would be seen through a telescope by the human eye. Some images are in black and white. A fuzzy or blurry image is shown only if there was no better image available of the object at the time the cards were made.

5. **Teams enjoy free exploration of cards.** Arrange students into teams of 4–6. Pass out a set of cards to each team. Tell students to share the cards with their team members. Encourage them to look through the cards at all the different objects. Give them 5–10 minutes to freely explore the cards before moving on to the sorting activity.

Sorting the Solar System Cards

1. **An activity to focus on the images.** Tell the class they will now play a sorting game to familiarize themselves with the images. This game will also help them to start thinking about what the objects are. Explain the procedure:

   a. **Divide up the cards.** Each team will divide their cards among its team members so each student has about 6–9 cards. It does not matter who has which cards.

   b. **Each student silently sorts his or her cards.** Emphasize that there is no right or wrong way to sort the cards. Students can make two, three, or four groups with their cards—it doesn't matter. However, they must have some reason for cards to be in the same group; there must be something that all the cards in a group have in common. They should display their sorted groups so that each card within a group is visible.

   c. **Team members take turns guessing sorts.** Each team member should examine the groups of cards that other team members have made and see if they can guess the reasons used to sort the cards the way they did. After everyone has had a chance to guess, each student should reveal to the team how he or she sorted the cards.

   d. **Sort again if time allows.** If there is enough time, teams can redistribute their cards and repeat the sorting activity.

The Solar System card sets can be used in a variety of ways to enrich your students' learning about Solar System objects. Sessions 3.3 and 3.4 outline some card-sorting activities; you and your students may discover other possible uses for the card sets as well.

TEACHER CONSIDERATIONS

TEACHING NOTES
Be sure to give your students a few minutes to simply look through and admire their card images. While it may be tempting to skip this free-exploration step, allowing students to marvel at the object images is an excellent way to generate interest in the objects for later card-sorting activities and discussions.

ABOUT THE SOLAR SYSTEM CARDS
Cards do not indicate the names of the objects shown. This is done deliberately to encourage students to study and compare the objects using only the given data. Withholding object names prevents students from being prejudiced by knowledge attached to a particular object. For instance, if a student was looking at card #6 (Mars) and wondering whether the object is a planet or a moon, he or she would not be able to say: "It's Mars, so it's a planet." Instead, the student would have to compare the object's size and appearance with other objects to come to a conclusion. The student might also have to consider the following question: "What else do I need to know about this object in order to decide whether it's a planet or not?"

The numbered order of the images is roughly the order in which the objects were discovered. Do not reveal this to the students until they have had some time to explore the cards and wonder for themselves if the numerical order of the cards has any significance.

The images shown were all taken at visible light wavelengths, which our eyes can see. Many published astronomical images are highly enhanced or taken at wavelengths of light that are invisible to the human eye. All of the card images were taken at visible wavelengths, with some adjustments made to the image brightness, contrast, and color, as is customary with astronomical photos. The image of the Sun (#2) is the least true to what our eyes would actually see—there is no way to reproduce the brightness of the Sun with ink on paper. (Even if it were possible, it would damage your eyes to look at such an image!) The image on the card was made using a special red filter that emphasizes solar surface features and prominences (projections of gas) around the edge of the Sun.

The quality of the images can be used to make inferences about the objects themselves. All of the images shown were the best ones available at the time the cards were manufactured. If an image is fuzzy or pixilated, this may be an indication that the object is too distant for a probe to approach it and obtain a sharper, clearer picture.

continued on page 421

Students may not understand why they are sorting the cards. Let them know that by examining and classifying images without knowing anything about them, they are doing what astronomers sometimes do when a new object is discovered.

SESSION 3.3 Exploring Diversity in the Solar System

Optional: Have teams use the interactive sorting program on the CD-ROM to show their sorts to the class.

2. Teams share their sorts with the class. Ten minutes before the session ends, regain the attention of the entire class. Ask for different teams to share some of their sorts with the class.

3. Objects can be classified in many ways. Point out that even though all teams used the same set of 36 cards, different teams were able to create different groups, or categories, with their cards. Explain that a *category* is a group of things or objects that have something in common with one another. Tell students that, as they have seen with this sorting activity, sometimes an object can fit into more than one category.

4. Categorizing something is not necessarily easy. Tell students that it can be difficult to categorize an object. Sometimes, scientists have difficulty deciding which categories to use and which objects to put into which categories.

5. Students will receive more information about the objects later. It is likely that some students will have tried to sort their cards by the kinds of objects the images appear to be, such as a moon, a planet, a comet, an asteroid, etc. Tell them that they will need more information about the objects in order to sort their cards accurately in this way. They will get more information in future sessions.

6. There are many different objects in the Solar System. Tell students that the cards they were looking at and sorting today represent only a small number of the many objects in the Solar System. Post on the concept wall, under Key Space Science Concepts:

Many diverse objects make up the Solar System.

7. Cards will be used again in future sessions. Ask teams to collect all 36 of their cards and put their set back into the sandwich bag or envelope in which it came. Let students know that they will have a chance to use the cards again and to learn more about the objects shown on the cards in upcoming sessions.

TEACHER CONSIDERATIONS

ABOUT THE SOLAR SYSTEM CARDS
continued from page 419

How the image was obtained may be a potential source of information. Many of the images were taken by the Hubble Space Telescope, which orbits the Earth. Others were taken by space probes sent to the object or by ground-based telescopes. It's important to note that you cannot necessarily infer an object's distance from Earth based on the quality of its image alone. For example, a probe that makes a fly-by of an object can get a clear image of a small object that is very distant from Earth. Consider also that while the Moon (#3) can be photographed clearly with a small ground-based telescope (given its close proximity to Earth), the image of Eris (#35) is fuzzy, despite the fact that it was photographed by one of Earth's most powerful ground-based telescopes at the Keck Observatory. The fuzziness of Eris' image could be an indication of just how very far away Eris is.

PROVIDING MORE EXPERIENCE
If time and interest allow, you could suggest some possible card-sorting variations to the class:

- Have one pair of students on a team sort all the cards. The rest of the team then tries to determine what criteria were used to sort the cards.
- Have everyone on a team work together to sort all the cards. Another team then tries to determine how the cards were sorted.

ASSESSMENT OPPORTUNITY
QUICK CHECK FOR UNDERSTANDING: DIVERSITY OF OBJECTS IN THE SOLAR SYSTEM
By the end of this session, students should understand that a large diversity of objects populate the Solar System, and that these objects can be categorized in different ways.

If you have students sort the Solar System cards over more than one class session, they can label their envelopes with their names and store the cards in groups using paper clips.

SESSION 3.4

Categorizing Objects in the Solar System

Overview

Objects are categorized according to certain characteristics they have in common with one another. Most students will know the names of the planets but not necessarily *why* these objects are considered planets. The decision in August 2006 by the International Astronomical Union to strip Pluto of its planetary status brings up the question: "How are Solar System objects categorized?" The continuing debate about Pluto among astronomers suggests that categorizing an object is not necessarily an easy or straightforward task. In this session, student groups work with the Solar System Cards to identify objects they think might be planets. Teams share the criteria they use to decide whether an object is a planet or not, and the class generates a list of potential planetary characteristics. The class then modifies the list through a guided, whole-class examination of the cards, which reveals additional information about some of the objects. After working through categorizations of various Solar System objects (planets, moons, asteroids, comets, and Kuiper Belt Objects), the class debates the question astronomers are still debating: "Is Pluto a planet?" During this session, the key concepts that will be added to the classroom concept wall are:

- *Scientists categorize objects in the Solar System by characteristics such as: shape and appearance, what they orbit, how large they are, and how far away their orbits are from the Sun.*
- *Not every Solar System object can be easily categorized.*

Categorizing Objects in the Solar System	Estimated Time
What Makes an Object a Planet?	15 minutes
Modifying the Class List of Planetary Characteristics	20 minutes
Categorizing Pluto	10 minutes
Total	**45 minutes**

What You Need

For the class:
- ❏ (optional) computer with large-screen monitor or LCD projector
- ❏ (optional) Space Science Sequence CD-ROM
- ❏ prepared key concept sheets from the copymaster packet or CD-ROM file
- ❏ a marker
- ❏ a piece of butcher paper
- ❏ Key for the Solar System Cards from Session 3.3

Unit Goals

The Solar System is centered around the Sun, the only star in the Solar System.

A wide variety of objects orbit the Sun in the Solar System.

Scientists categorize Solar System objects according to their characteristics; however, not all objects can be easily categorized.

Objects in the Solar System are in regular and predictable motion.

The Solar System is mostly empty space, and is very large compared to the objects located within it.

TEACHER CONSIDERATIONS

Key Vocabulary

Scientific Inquiry Vocabulary

Category
Characteristic
Evidence
Model
Observation
Prediction
Scale
Scale model
Scientific explanation

Space Science Vocabulary

Asteroid
Astronomical Unit (AU)
Comet
Diameter
Heliosphere
Kuiper Belt Object (KBO)
Moon
Orbit
Planet
Sphere
Star
System

SESSION 3.4 Categorizing Objects in the Solar System

For each team of 4–6 students:
- ❑ 1 set of 36 Solar System Cards from Session 3.3
- ❑ 1 "Planet" category place card (see Getting Ready)
- ❑ 1 "Not a Planet" category place card (see Getting Ready)
- ❑ 1 "Don't Know/Unsure" category place card (see Getting Ready)

Getting Ready

1. **Prepare the key concept sheets.** Make a copy of each key concept and have them ready to post onto the classroom concept wall during the session.

2. **Prepare the class chart.** Using a marker, write across the top of a piece of butcher paper: What Makes an Object a Planet? Set this aside for later, when the class begins discussing how they chose to designate an object as a planet. **Important Note:** At the end of the session, store this list away for use again in Session 3.11.

3. **Decide how you will divide the class into teams of 4–6 students.**

4. **Prepare category place cards for each team.** Each team will need a set of three different category place cards. Consider using different color sheets of paper to distinguish between the three categories. Using a marker, write the word "Planet" across the top of a sheet of paper. Do the same for "Not a Planet" and "Don't Know/Unsure." Make enough of these category place cards for each team to have a set; put a set with each set of Solar System cards.

5. **Optional: If you plan to use the Space Science Sequence CD-ROM, set up the computer with a large-screen monitor or LCD projector.** The CD-ROM has a useful interactive activity—The Solar System Card Sort—which you may use to lead the class-sorting activity and discussion. You may also have teams show their sorts to the class using the CD-ROM. Instructions for using the CD-ROM can be found in Session 3.3.

TEACHER CONSIDERATIONS

TEACHING NOTES

The key concepts can be posted in many different ways. If you don't want to use sentence sheets, here are some alternatives:

- Write the key concepts out on sentence strips.
- Write the key concepts out before class on a posted piece of butcher paper. Cover each concept with a strip of butcher paper and reveal each one as it is brought up in the class discussion.

SESSION 3.4 Categorizing Objects in the Solar System

🢂 What Makes an Object a Planet?

1. **Reintroduce the cards.** Tell students that today they will be working with the Solar System cards again. Remind them that the images on the cards depict various objects within the Solar System. Refer to the posted key concept from Session 3.3:

   Many diverse objects make up the Solar System.

2. **More information about the objects.** Tell students that during this session they will learn more about some of the different objects on the cards.

3. **What defines a planet?** Students will use what they learn about the different objects to answer the following question: "What makes an object a planet?"

4. **If necessary, remind students of the definition of the term** *category*. A *category* is a group of things or objects that have something in common with one another. Remind students that in Session 3.3 they took turns sorting card objects into different categories and had their teammates guess the objects.

5. **Categorizing the Sun.** Ask students to decide how many of the cards depict objects that are stars. [Only one. There is only one star in the Solar System—the Sun.] If necessary, remind students of the key concept from Session 3.2:

   The Solar System is centered around the Sun—the only star in the Solar System.

6. **Teams work to put cards into categories.** Tell students they will be working with the cards in teams of 4–6. Let them know their first task will be to find the Sun card (#2) and set it apart. Since it is the only star in the Solar System, it is in a category by itself. Then each team should work together to divide the rest of their cards into the following three categories:

   - Planet: Objects they think are planets.
   - Not a Planet: Objects they think are not planets.
   - Don't Know/Unsure: Objects they are not sure of or cannot agree about.

   Explain that teams will put cards for each category on the appropriately labeled place card. The labeled place cards will help students to visually keep track of the category into which they are putting objects.

Some students will be very uncomfortable with objects they cannot easily categorize. Emphasize to your students that it's okay to be unsure!

TEACHER CONSIDERATIONS

SESSION 3.4 Categorizing Objects in the Solar System

7. **Teams should have reasons for their sorts.** Tell students that you will be asking them to explain why they chose to categorize their cards the way they did.

8. **Divide the class into teams of 4–6 students.** Pass out a set of cards and category place cards to each team. Allow five minutes for teams to organize their cards into the three categories. Circulate among teams, paying special attention to how they decide to categorize an object as a planet.

9. **Optional: Teams show their sorts using the CD-ROM.** If you have extra class time and would like to do so, have a few teams show their sorts to the class using the CD-ROM's interactive card-sorting program. Click on The Solar System Card Sort activity. As teams demonstrate their sorts to the class, ask them to share what characteristics they used to designate an object as a planet. Using their answers, lead the discussion detailed below in Step #10.

10. **Sharing reasons for categorizing objects as planets.** After teams have finished sorting their cards, regain the attention of the class. Ask teams to share what characteristics they used to categorize an object as a planet. List their reasons on the piece of butcher paper titled: What Makes an Object a Planet? Listed below are some reasons students might give—but not all of them are correct!

   - They recognize the object and know it is a planet. (Note: Tell students they should try to categorize the objects based on the information shown on the cards alone, not on any previous knowledge they may have of the object.)
   - It looks round or spherical.
   - It is bigger than a certain size. (Note: Ask what range of sizes they think is right for a planet.)
   - It appears to have an atmosphere.
   - It orbits the Sun. (Note: Acknowledge that this is important, but ask if the card gives them this information. [No. The card does not have any information about *what* the object orbits around.])

Modifying the Class List of Planetary Characteristics

Confirm that it is difficult to sort planets from other objects without more information. Tell the class that now you are going to reveal some additional information that scientists have gathered about some of the objects. Using this information, the class will add to or delete from the list of planetary characteristics they currently have. They may also want to move cards to different categories based on this new information.

TEACHER CONSIDERATIONS

TEACHING NOTES
It's okay if not all the reasons initially listed are correct. As the activity continues, the class will be modifying their list of planetary characteristics.

SESSION 3.4 Categorizing Objects in the Solar System

The vast diversity of objects in the Solar System can be a point of great interest for students. If you have additional classroom time, consider going through the slightly more extensive Revealing More About Comets and Asteroids card-sorting activity detailed under Teacher Considerations, pages 431. This extended discussion takes more time, but students will have a chance to look at more Solar System objects.

Revealing Asteroids and Comets

1. **Have students find #23 (Ida) and #26 (nucleus of Comet Borrelly). Do not reveal the object names.** Call on students to describe ways in which these objects are either similar or different. [They are similar in shape but different in size.]

2. **Tell students that both objects orbit the Sun.** Ask if anyone thought these objects might be planets. Chances are that most students thought they were too small or the wrong shape to be planets. Add "size" and "spherical shape" to the class list of planetary characteristics if they are not already there. Confirm that #23 and #26 are not planets, even though they both orbit the Sun.

3. **Reveal that #23 is Ida, an asteroid.** Ida orbits the Sun as part of the Asteroid Belt. Located between Mars and Jupiter, the Asteroid Belt is a collection of many multi-sized chunks of rock that orbit the Sun.

4. **Reveal that #26 is the nucleus of Comet Borrelly.** Like asteroids, comets can also orbit the Sun. The nucleus of a comet is often referred to as a "dirty snowball" since it is a mixture of ice, gases, and rocky dust. If, in its orbit, the comet gets close enough to the Sun for the ice to vaporize, a *coma*, or glowing cloud (sometimes called the *head* of the comet) can form around the nucleus. Comets close enough to the Sun will also have tails stretching away from the Sun. Most comets, though, are too far from the Sun to have tails. Comets are thought to have originated from two distinct regions in the Solar System: the Kuiper Belt and the Oort Cloud. (See the Background Information for Teachers section for more information about these two regions.)

5. **Several other cards show comets and asteroids.** Tell students that they can probably recognize more comets and asteroids, although some asteroids look like planets and some moons look like asteroids. Tell students that after you sort out the planets and moons, the remaining objects will be mostly comets and asteroids. It is not necessary to take the time to sort the comets and asteroids, but if students are interested, there are optional steps for exploring these objects. A summary list of which objects are comets and which are asteroids is provided in the Teacher Considerations on page 433.

6. **Confirm that although asteroids and comets orbit the Sun, they are not considered planets.**

TEACHER CONSIDERATIONS

SCIENCE NOTES
Defining *Asteroids* and *Comets*

Asteroid: Any of the thousands of small rocky objects that orbit around the Sun. Although most asteroids are located between the orbits of Mars and Jupiter in the Asteroid Belt, some pass closer to the Sun than Earth does, and others have orbits that take them well beyond Jupiter. The largest asteroid known is Ceres. With a diameter of 933 km, it's about as wide as the state of Texas!

Comet: A small chunk of ice, gas, dust, and rocky material that travels in an elliptical orbit around the Sun. A comet is made up of a nucleus, coma, and tail(s). The nucleus of the comet, which consists of ice, gases, and rocky dust, is often referred to as a "dirty snowball." A coma forms around the nucleus when the ice in the nucleus is vaporized by the Sun. If the comet travels close enough to the Sun, it may form a dust tail and/or a gas ion tail. These tails are driven off the comet's nucleus by the Sun's energy and generally point away from the Sun, regardless of the direction in which the comet is moving.

PROVIDING MORE EXPERIENCE
Revealing More About Comets and Asteroids

1. **Ask students to find a card that shows a full comet with its coma and tail.** Students should find #32 (Comet Hale-Bopp). Comet Hale-Bopp was one of the brightest comets of the 20th century. You can easily see two parts of its tail. One part is a stream of dust particles, and the other is a stream of glowing gas. Have students put this card with Comet Borrelly (#26) to start a "Comet" pile in the "Not a Planet" category.

2. **Next, ask students to find a card that shows a comet that has broken into pieces.** Students should find #30 (Comet Shoemaker-Levy 9). This comet was attracted by the gravity of Jupiter and went into orbit around Jupiter. When it was discovered in 1993, it had recently broken up. The fragments smashed into Jupiter less than a year later, and the comet no longer exists. Have students place card #30 in the group with the other comets in the "Not a Planet" category.

3. **Have students find #9.** Card #9 shows the nucleus of Comet Halley, one of the most famous of all comets. This is the first comet that an astronomer (Edmund Halley) predicted would return, and it is the first comet to have a photograph taken of its nucleus. It comes around in its orbit about every 76 years. The next time it comes around will be in 2062. Have students place card #9 in the group with the other comets in the "Not a Planet" category.

4. **Have students find #21 and #29.** Card #21 shows the nucleus of Comet Tempel 1, which was the target of the Deep Impact mission. This mission had a probe that crashed into the comet nucleus to help astronomers learn more about its structure

continued on page 433

SESSION 3.4   Categorizing Objects in the Solar System

If necessary, remind your students of the following key concepts from Session 3.2: Planets orbit stars. Moons orbit planets.

Revealing Moons and Planets

1. **Have teams find #8 (Mercury) and #11 (Callisto).** Tell the class that one of these objects is considered a planet and the other a moon, even though they are similar in size and appearance to one another (both are large and spherical). Ask students what they would need to know to decide which object is the planet. [They would need to know what each object orbits around. If the object orbits the Sun, it is a planet; if the object orbits a planet, it is a moon.] If "orbits the Sun" is not listed as a planetary characteristic, add it to the list.

2. **Reveal that #8 is the planet Mercury, and that #11 is Callisto, a moon of Jupiter.** Tell students that Mercury is the innermost planet orbiting the Sun. Remind students that Callisto is one of the moons of Jupiter they observed in Session 3.2. Give teams a chance to make sure #8 (Mercury) is on their "Planet" place card. Have them put #11 (Callisto) on the "Not a Planet" place card, in a separate pile for moons.

3. **Remind students that moons are part of systems that orbit the Sun.** Tell students that even though we say that Callisto orbits Jupiter, Jupiter orbits the Sun, taking all its moons along. Callisto is part of a system that orbits the Sun—the center of the Solar System.

4. **Reveal the card numbers of the other Galilean moons.** Have teams put #10 (Ganymede), #12 (Io), and #13 (Europa) with #11 (Callisto) in the moon pile on their "Not a Planet" place card. Remind students that #10 (Ganymede), #12 (Io), and #13 (Europa) are the other three moons of Jupiter, which they observed in Session 3.2.

5. **Have students find #4 and #14.** Tell them that both #4 and #14 have cloudy atmospheres, but that one is Venus, a planet, and the other is Titan, the largest moon of Saturn. Ask students which one they think is Venus. [#4 is Venus; Titan is much smaller.] Have students put #4 (Venus) in their planet pile and #14 (Titan) in their moon pile.

6. **Remove "atmosphere" from the list of planetary characteristics.** Be sure to cross off "atmosphere" if it is listed under What Makes an Object a Planet? Tell students that a few other moons have very thin atmospheres, while Mercury, a planet, has almost no atmosphere!

TEACHER CONSIDERATIONS

Revealing More About Comets and Asteroids
continued from page 431

and composition. Card #29 shows the nucleus of comet Wild 2 (pronounced *Vilt*). In 2003, the Stardust probe flew near Comet Wild to capture particles floating in space. The particles were returned to Earth to be studied. Have students place cards #21 and #29 in the group with the other comets in the "Not a Planet" category.

5. **Have students find #24 and #33.** Card #24 shows an asteroid called Eros. Card #33 shows an asteroid called Itokawa (ee-toe-ka-wa). These asteroids are not part of the Asteroid Belt. They have orbits that carry them closer to Earth's orbit. There have been missions by space probes to both of these asteroids. Have students put these cards with Ida (#23) to start an "Asteroid" pile in the "Not a Planet" category.

6. **Have students find #22 (Phobos).** Phobos has the size and appearance of an asteroid, but Phobos orbits Mars, so it is categorized as a moon. Mars orbits the Sun near the inner edge of the Asteroid Belt, and Phobos may once have been an asteroid that was "captured" into orbit by the gravity of Mars.

7. **Considering #31 (Dactyl).** Have students find card #31 (Dactyl). Tell students that since Dactyl orbits the asteroid Ida (#23), it is often called a moon of an asteroid. Let students know that Dactyl was the first moon of an asteroid ever discovered.

8. **Is Dactyl a moon or an asteroid?** Have teams discuss whether Dactyl should be categorized as a moon or an asteroid. Survey the class and call on a few students to explain why they classified Dactyl the way they did.

9. **Confirm that although asteroids and comets orbit the Sun, they are not considered planets.**

10. **Have students sort the remaining asteroids and comets.** Here is a summary of the cards that show comets and asteroids.
    Comets:
    #9 nucleus of Comet Halley
    #21 nucleus of Comet Tempel 1
    #26 nucleus of Comet Borrelly
    #29 nucleus of Comet Wild 2 (Wild is pronounced *Vilt*)
    #30 fragments of Comet Shoemaker-Levy 9 (before crashing into Jupiter)
    #32 Comet Hale-Bopp
    Asteroids:
    #17 Ceres (The largest asteroid, it was given "dwarf planet" status in 2006.)
    #23 Ida
    #24 Eros
    #31 Dactyl (It orbits Ida, so some call it a moon of an asteroid.)
    #33 Itokawa

SESSION 3.4 Categorizing Objects in the Solar System

7. Review the What Makes an Object a Planet? list. Ask students what planetary characteristics remain on the list. [Large, spherical, and orbits the Sun.] Ask students what cards they would classify as planets, based on these characteristics. Give them a few minutes to organize their cards and then say that scientists using these criteria would definitely categorize the following eight objects as planets:

   #8   Mercury
   #4   Venus
   #1   Earth
   #6   Mars
   #5   Jupiter
   #7   Saturn
   #16 Uranus
   #18 Neptune

Students may ask why Pluto has been omitted from the list. Tell them that in 2006, at a meeting of The International Astronomical Union, several hundred astronomers voted for a new definition of a planet. Under this new definition, Pluto did not qualify for planetary status. Not all astronomers agreed with this decision, and there is still plenty of debate among the astronomical community (and others) about Pluto. Tell students that they'll soon have a chance to discuss whether or not Pluto should be considered a planet!

8. Objects astronomers would definitely categorize as moons. Ask students which objects they think are moons. Remind students of the key concept from Session 3.2:

   Moons orbit planets.

Tell the class which planet each moon orbits as you reveal the card numbers of the following eleven moons:

   #3   The Moon (orbits Earth)
   #10 Ganymede (orbits Jupiter)
   #11 Callisto (orbits Jupiter)
   #12 Io (orbits Jupiter)
   #13 Europa (orbits Jupiter)
   #14 Titan (orbits Saturn)
   #15 Tethys (orbits Saturn)
   #19 Triton (orbits Neptune)
   #20 Ariel (orbits Uranus)
   #22 Phobos (orbits Mars)
   #25 Phoebe (orbits Saturn)

TEACHER CONSIDERATIONS

PROVIDING MORE EXPERIENCE

If time and interest allow, go through all the moons and their planets using the Key for the Solar System Cards or using the cards with the names of the objects. If you do not have time to do this, make sure students understand that this list is only a partial list of all the moons in the Solar System.

TEACHING NOTES

Here is some background about Pluto and the definition of a planet, which may help you address student questions. Simply presenting this information to your students may indicate that there is no further reason to consider the question of which objects in the Solar System are planets. By withholding the "official word" on the subject, you encourage students to explore the characteristics of Solar System objects and consider for themselves which categories make sense.

In August 2006, 424 members of the International Astronomical Union (IAU) voted on a resolution to set new criteria for an object in the Solar System to qualify as a planet. The resolution was approved by a vote of 237-157, with 30 people abstaining.

These are the criteria adopted by the IAU that qualifies an object as a planet in the Solar System:

a) It is in orbit around the Sun.
b) It has sufficient mass for its self-gravity to overcome rigid body forces so that it assumes a hydrostatic equilibrium shape. (That means that its gravity pulls it into a shape close to a sphere.)
c) It has cleared the neighborhood around its orbit.

Pluto does not satisfy (c). There are many known objects—and probably many more undiscovered ones—that have orbits in the neighborhood of Pluto, so Pluto does not qualify as a planet. The IAU had once considered calling Ceres (the largest asteroid) and Eris (a Kuiper Belt Object, which is bigger than Pluto) planets, but they were rejected for the same reason as Pluto. A new class called "dwarf planets" was instituted, which includes Pluto, Ceres, and Eris. Some other large asteroids and Kuiper Belt Objects are being considered for "dwarf planet" status.

Pluto's loss of planetary status hasn't been without controversy or protest! Despite the IAU's decision to demote Pluto in 2006, many astronomers would like to see Pluto reinstated as the ninth planet in the Solar System.

In response to Pluto's demotion, the American Dialect Society chose *plutoed* as its 2006 Word of the Year. The society defines *to pluto* as "to demote or devalue someone or something, as happened to the former planet Pluto when the General Assembly of the IAU decided Pluto no longer met its definition of a planet."

SESSION 3.4 Categorizing Objects in the Solar System

Categorizing Pluto

1. **The controversy surrounding Pluto.** Let students know that although Pluto has long been considered a planet in the Solar System, in 2006 astronomers voted in favor of a definition of a planet that does not include Pluto. Despite this decision, the debate continues among many astronomers as to whether or not Pluto should be called a planet. Tell students that they're about to discuss this question, but first you'll give them some more information to help them think through this issue.

2. **Have students find #17 (Ceres), #27 (Pluto), #28 (Charon), and #35 (Eris). Do not reveal any of the object names yet!** (Ceres, #17, will be in the asteroid group. The others may not have been categorized yet.) Ask students to describe the ways in which these objects appear similar to one another. [All the pictures of objects are not very clear, and the objects are all spherical.]

3. **Tell the class that some astronomers have considered adding all these objects to the planet category, but other astronomers did not agree.** Take a survey of the teams to see which of these objects, if any, they thought might be planets.

4. **Reveal that #17 is the asteroid Ceres.** Students may be surprised that it looks so planet-like. Ask them why they think astronomers do not consider it to be a planet. [It's too small. It is one of many thousands of asteroids that have similar orbits.] Have them put #17 (Ceres) in the asteroid pile on their "Not a Planet" place card.

5. **Reveal that #27 is Pluto, and #28 is Charon (one of Pluto's moons).** Ask students why they think Charon is not considered a planet. [Because it is smaller than Pluto, and it orbits Pluto.] Say that because Charon is more than half the diameter of Pluto, some astronomers have described Charon and Pluto as a double-dwarf planet system.

6. **Reveal that #35 is Eris, an object that is somewhat more distant than Pluto.** Discovered in 2003, Eris is similar to Pluto. It has a diameter of 2400 km and is slightly larger than Pluto.

7. **Introduce Kuiper Belt Objects (KBOs).** Tell students that, as of 2007, Eris is the largest-known object in a collection of objects beyond the orbit of Neptune called the Kuiper Belt. Like the Asteroid Belt, the Kuiper Belt is composed of multi-sized objects called Kuiper Belt Objects or KBOs. Pluto is considered by many astronomers to be a large KBO, like Eris.

If you have additional classroom time, consider examining other Kuiper Belt Objects. Optional steps are detailed under Teacher Considerations on page 437.

TEACHER CONSIDERATIONS

PROVIDING MORE EXPERIENCE
More KBOs and Objects Beyond Neptune
1. **Have students guess which other cards show KBOs.** Ask teams what evidence they used to make their guesses. Reveal that #34 (Quaoar, pronounced *KWAH-war*) is a KBO. Tell the class that #36 (Sedna) is so far away that astronomers are unsure whether it is a KBO or an object from a group located beyond the Kuiper Belt.
2. **Have students find #19 (Triton).** Triton will be with the moons since it orbits the planet Neptune. Triton is slightly larger than Pluto, but in some ways they are quite similar. Triton may have once been a KBO orbiting the Sun, as Pluto now does. However, now that Triton orbits Neptune, it is categorized as a moon.

SCIENCE NOTES
In 2005, the Hubble Space Telescope discovered two small moons orbiting Pluto, bringing the number of known satellites for Pluto to three. This is a good indication of space science as a constantly changing and very exciting field—as new discoveries are made, more missions are launched, and new instruments are invented and refined. Whenever possible during the unit and over the course of the sequence, encourage your students to obtain up-to-date astronomy and space-mission information by accessing resources such as NASA's website (www.nasa.gov).

At 14 billion miles, Eris is currently over two times farther from the Sun than Pluto. It is located in the Kuiper Belt, a swarm of icy objects orbiting the Sun beyond Neptune's orbit.

SESSION 3.4 Categorizing Objects in the Solar System

8. **Is Pluto a planet?** Ask students whether they now think Pluto should be categorized as a KBO or a planet or both. Follow up by asking whether Eris, and some or all of the KBOs, should be categorized as planets. Accept all opinions.

9. **More information may be needed.** Tell them that later they will get another chance to think about and discuss whether Pluto should be called a planet or not. Ask whether there is more information they would like to have that might help them categorize Pluto. Accept their answers. Tell them that they will be learning more about what planets are made of, how they are arranged, and how they move before they discuss Pluto again as a class.

10. **Astronomers characterize objects using certain characteristics.** Tell students that their object cards can be categorized in many different ways, depending on what object characteristics they choose to look at. (Remind students of the many different sorts they were able to make with the same cards in Session 3.3.) Say that astronomers categorize objects in the Solar System using certain characteristics. Also tell students that some objects can be difficult to categorize. Post on the concept wall, under Key Space Science Concepts:

    Scientists categorize objects in the Solar System by characteristics such as: shape and appearance, what they orbit, how large they are, and how far away their orbits are from the Sun.

    Not every Solar System object can be easily categorized.

11. **Collect the Solar System card sets.** Have each team put their card sets into the bag or envelope in which they came. Collect the sets and keep the class list of What Makes an Object a Planet? for use in Session 3.11.

TEACHER CONSIDERATIONS

SESSION 3.5

Researching Objects in the Solar System

Overview

In previous sessions with the Solar System card sets, students explored the great diversity of objects found within the Solar System. They also used characteristics such as size and appearance to help them categorize objects. In this session, students learn more about an assigned Solar System object by conducting their own research. Working in pairs, students create interesting and creative learning stations to share the results of their research with one another. Each learning station includes a color illustration and scale model of the assigned object, as well as a set of creative travel brochures sharing fun and accurate information about the object. Students prepare these projects for Session 3.7, when the class will "tour" the Solar System by visiting all of the learning stations to find out more about the different characteristics of various Solar System objects. There are no new key concepts for this session.

Researching Objects in the Solar System	Estimated Time
Learning Station and Travel Brochure Guidelines	20 minutes
Researching Assigned Objects, Creating Scale Models, and Designing Brochures	25 minutes
Total	45 minutes

What You Need

For the class:
- ❏ 1 copy of the Assignment Sheet for Solar System Learning Stations from the copymaster packet or the CD-ROM file
- ❏ 1 set of Solar System Object fact sheets from the copymaster packet or the CD-ROM file
- ❏ (optional) sample travel brochures
- ❏ assorted colors of modeling clay (for scale models of small planets and moons)
- ❏ newspaper (for scale models of larger planets)
- ❏ 1 roll of masking tape (for scale models of larger planets)
- ❏ some chalk dust or flour (for scale models of the Kuiper Belt and Oort Cloud)
- ❏ a few pairs of scissors
- ❏ a few rulers (marked in centimeters)
- ❏ 16 glue sticks
- ❏ colored markers and/or colored pencils
- ❏ construction paper in different colors
- ❏ (optional) various research sources students can access

For each pair of students:
- ❏ 1 copy of the Solar System Learning Station Guidelines student sheet from the copymaster packet or CD-ROM file

Unit Goals

The Solar System is centered around the Sun, the only star in the Solar System.

A wide variety of objects orbit the Sun in the Solar System.

Scientists categorize Solar System objects according to their characteristics; however, not all objects can be easily categorized.

Objects in the Solar System are in regular and predictable motion.

The Solar System is mostly empty space, and is very large compared to the objects located within it.

TEACHER CONSIDERATIONS

TEACHING NOTES

This session, as well as Sessions 3.6 and 3.7, have proven to be extremely popular with students! Be prepared to allot additional class time for these sessions if you can.

Other options for storing learning-station items include large plastic storage tubs or baskets. Anything that will allow students to store their works-in-progress is fine. Large, black, plastic trash bags, however, can do double-duty—as storage containers and as display backgrounds of "space" for the learning stations.

If you have additional class time or an extended period, you may want to combine this session with the next one. (Students complete their travel brochure designs in Session 3.6.)

Key Vocabulary

Scientific Inquiry Vocabulary

Category
Characteristic
Evidence
Model
Observation
Prediction
Scale
Scale model
Scientific explanation

Space Science Vocabulary

Asteroid
Astronomical Unit (AU)
Comet
Diameter
Heliosphere
Kuiper Belt Object (KBO)
Moon
Orbit
Planet
Sphere
Star
System

SESSION 3.5 Researching Objects in the Solar System

- ❏ 2 copies of their assigned object's fact sheet from copymaster packet or CD-ROM file
- ❏ 1 large, black, plastic trash bag (to store and then display learning-station items)

For each student:
- ❏ 1 copy of the Solar System Travel Brochure Guidelines student sheet from copymaster packet or CD-ROM file
- ❏ a piece of blank white 9"x 12" construction paper for the travel brochure
- ❏ a pencil

Getting Ready

1. **Decide student assignments for objects listed on the Assignment Sheet for Solar System Learning Stations.** Assign a pair of students to research each Solar System object. Record students' names on the assignment sheet. Note that there are 16 objects listed. See Teacher Considerations on page 443 for assignment suggestions if you have fewer than or more than 32 students in your class.

2. **Make several copies of each assigned object's fact sheet.** You'll need a complete set of fact sheets for your own reference. Each team of students should also have one copy per student of their assigned object's fact sheet in order to conduct their research.

3. **Make a copy of the Solar System Learning Station Guidelines student sheet for each pair of students.**

4. **Make a copy of the Solar System Travel Brochure Guidelines student sheet for each student.**

5. **Have all art materials organized and ready for use in an easily accessible location in the classroom.**

6. **Make a scale model of Earth.** Do this by rolling modeling clay into a ball close to 13mm in diameter. You will use this as an example of a scale model in this session, as well as with scale models that the students make in Session 3.8.

7. **Have any research sources available for your students to access.** Besides the object fact sheets, students should obtain information about their object using other resources if possible. If you have any additional resources available (such as a computer with Internet access, a book about the Solar System, or an encyclopedia), decide where in the classroom you will place these materials. If you do not have access to other resources, consider assigning additional research as homework for your students.

TEACHER CONSIDERATIONS

TEACHING NOTES

If you have fewer than 32 students: Choose the objects that interest you most and eliminate objects that do not have students to cover them. (Note for Session 3.7: Be sure to cross out the objects that are not represented on the Tour of the Solar System—Travel Notes and Tour of the Solar System—Comparing Object Characteristics student sheets before making copies for the class.) Alternatively, you may assign learning stations as solo projects for any of your advanced or highly independent students. This option eliminates teamwork for the student but allows for more objects to be represented during the class tour of the Solar System in Session 3.7.

If you have more than 32 students: Consider assigning teams of three students to research each object. Alternatively, you could choose additional objects for student pairs to research from the Key for the Solar System Cards on page 417 from Session 3.3 or assign Earth as one of the objects. (Note: These additional objects will not have prepared fact sheets. Student teams will need to conduct their research using their own resources. This might be a good option for advanced or highly independent and motivated students!) If your class is researching extra objects, consider making these objects optional stops on the tour in Session 3.7 or extend the tour beyond a single session.

If students do not have access to additional resources at home, consider setting aside some time for the class to visit the school library so they can obtain more research sources.

Note about scale model materials. Each student team will construct a scale model of their object using the information listed at the top of the object's fact sheet. The clay should be used by students making scale models of the moons and smaller planets. Students making models of the larger planets—Jupiter, Saturn, Uranus, and Neptune—can ball up pieces of newspaper to the approximate size needed and form a sphere by wrapping masking tape around the newspaper a few times. Students modeling the Asteroid Belt, Kuiper Belt, or Oort Cloud should use either chalk dust or flour. The construction of the learning station requires only a smear of dust that can be obtained from a chalkboard eraser. (If you know a student who is allergic to dust or for whom dust might trigger asthma, assign that student to one of the larger Solar System objects.)

ASSIGNMENT SHEET FOR SOLAR SYSTEM LEARNING STATIONS

Solar System Object	Students Assigned to Research Object
Mercury	
Venus	
The Moon (Luna) – Moon of Earth	
Mars	
Asteroid Belt	
Jupiter	
Io – Moon of Jupiter	
Europa – Moon of Jupiter	
Callisto – Moon of Jupiter	
Saturn	
Titan – Moon of Saturn	
Uranus	
Neptune	
Pluto	
Kuiper Belt	
Oort Cloud – Cold Distant Comets	

Teacher Reference Sheet—Space Science Sequence 3.5

Encourage your students to be as creative and imaginative as possible when designing their learning stations. This session and Session 3.6 are both excellent opportunities for students to study a Solar System object in an enriching and engaging way.

SESSION 3.5 Researching Objects in the Solar System

8. Decide how and where you would like students to store their learning station projects. You may choose to have students place their works-in-progress in large trash bags or storage bins. Designate a location in your classroom for students to place their storage containers.

GO! Learning Station and Travel Brochure Guidelines

1. Student teams will create learning stations for Solar System objects. Remind students of the many Solar System object images they studied while using the card sets in previous sessions. Tell students that today they will be working in teams to research and learn more about just one of these objects. They will share their research with the rest of the class by creating a learning station. Tell students that they should design their learning stations so that they can share interesting and accurate information about their object in a creative and fun way.

2. Students will "tour" the Solar System by visiting learning stations. Tell students that they will learn more about different Solar System objects by visiting each team's learning station. The class will not be "touring" the Solar System today—they will do this in Session 3.7, once all the learning stations are ready.

3. Explain what each learning station should include. Each learning station should include the following elements:

- A three-dimensional scale model of the object.
- A sign or banner clearly stating the name of the object.
- A color illustration of the object.
- A travel brochure from each member of the team, which shares interesting information about the object.

TEACHER CONSIDERATIONS

One teacher said, "Students really enjoyed researching an object in the Solar System. Many went beyond the requirements. Some made and painted their own scale models for the larger planets from styrofoam balls. One pair made a question/answer quiz with their presentation; several others developed raps. One made a PowerPoint that was impressive for a sixth grader."

Another said, "For English Language Learners, the graphics options offer strong opportunities for language and literacy acquisition."

SESSION 3.5 Researching Objects in the Solar System

A scale model is a representation of an object that is smaller or larger than the actual size of the real object. Every part of the real object is measured and made smaller or larger by the same factor to create the scale model.

4. **Scale models have to be completed first!** Tell students that their first priority should be to create a scale model of their object. The fact sheet they will get states what size they should make their model. After they make their models, they should design a sign or banner for their station and begin work on a color illustration of their object. Only after they have completed these learning station elements should they begin work on their travel brochures! Tell them that in the next session, they will have the full period to work on finishing up their brochures.

5. **Teams will research objects using fact sheets and other sources.** Review the format of the fact sheets with the class. Suggest other sources for students to look into when conducting their research. Make sure students understand that although they may be creative in designing their learning stations, the information they share should be **accurate** and **correct.** Challenge and encourage students to put energy and creativity into designing their stations so that the tour will be both fun and educational for the class.

6. **The fact sheet shares information for constructing a scale model.** Information for constructing their object's scale model is given at the top of each fact sheet. (If necessary, review what *diameter* means with the class.) Explain that their model must be sized correctly, since each object's scale model will be used later on in another activity where the scale will matter.

7. **Constructing object scale models.** Tell the class that teams will be making scale models of moons and planets. Students making models of the larger planets—Jupiter, Saturn, Uranus, and Neptune—should crumple newspaper into balls and use masking tape to hold the newspaper in a spherical shape. Smaller moons and planets will be made of modeling clay. Show them the 13mm diameter model of Earth that you made. Suggest that teams tape small models like this onto a piece of paper, so that they won't get lost. However, the models should still be detachable. Students making models of the Asteroid Belt, Kuiper Belt, and Oort Cloud should use chalk dust or flour taped or glued to paper.

8. **Discuss the travel brochures.** Stress to the class that the travel brochures will be the most unique part of each of their learning stations. Good travel brochures will make the class tour of the Solar System much more interesting. Each student should design a brochure—one that is creative, but also **factually correct.**

TEACHER CONSIDERATIONS

TEACHING NOTES
Clearly emphasize that students should complete their scale models before moving on to the travel brochures. The scale models are necessary for Sessions 3.7, 3.8, and 3.9.

SESSION 3.5 Researching Objects in the Solar System

9. **Travel brochures that are accurate and also creative.** Tell the class that an interesting travel brochure doesn't just list facts. Use the example below about Chile's Atacama Desert to illustrate this to the class.

A factually accurate but uninteresting brochure:
About the Atacama Desert in Chile:
- Rainfall has never been recorded in some places in the Atacama Desert.
- The Atacama Desert has many cloudless days.
- Located in the Southern Hemisphere, the Atacama Desert has seasons opposite to those in the Northern Hemisphere.
- The human population in the Atacama Desert is 202,259 in a region of 29,000 square miles.

A factually accurate and more interesting brochure:
Come Visit the Atacama Desert in Chile!
- Do you have athlete's foot or other fungi growing on your body? Then come to the Atacama Desert. It's so dry that there are places in it where rainfall has never been recorded. Those pesky fungi of yours will just dry up and wither away!
- Do you love sunny days? Do clouds and rain get you down? Come to the Atacama Desert where it is almost always sunny. You're sure to find a trip here a big mood booster! Cloudless nights also make for many memorable evenings out stargazing.
- Tired of those cold January days? Come to the Atacama Desert where January is summer and one of our hottest months of the year.
- Crowds driving you crazy? Hate waiting in line at the supermarket? You're sure to smile when you come to the Atacama Desert, where you're quite likely to see no one. There is an average of only 7 people per square mile!

10. **Both brochures share the same facts but have different impacts on the reader.** Ask the class which brochure description they enjoyed listening to more. Tell them that they should each try to design a brochure as interesting and fun to read as the one you just read. Show the class the white paper they should use for the brochures and suggest that they design their brochures to fold in three parts. (Demonstrate this to the class if necessary.)

11. **Optional: Show the class some examples of good travel brochures.**

TEACHER CONSIDERATIONS

TEACHING NOTES
Consider placing the sample travel brochures with the craft supplies and additional research sources for students to look over.

Another Option: Show the class travel brochures from a travel agency.

SESSION 3.5 Researching Objects in the Solar System

SOLAR SYSTEM LEARNING STATION GUIDELINES

Learning Station Project Checklist
Be sure to prepare the following items for your station:
- **Title Sign or Banner**—Make a sign or banner that clearly displays your object's name.
- **Drawing of Object**—Create a color illustration of your object.
- **Accurate Scale Model**—Make a scale model of your object. Refer to your object fact sheet to find out how you should make your model.
  The most important thing is that your model is sized correctly!
  Although the final model doesn't have to look exactly like your object, if you would like to decorate your model to make it look more realistic, go ahead!
- **Travel Brochures**—Each team member should prepare a travel brochure sharing interesting and accurate information about the object.

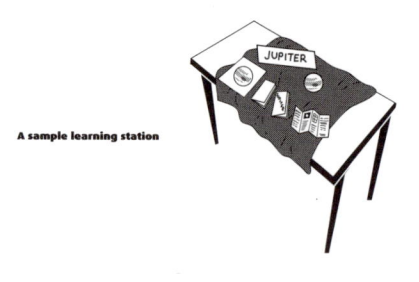

A sample learning station

Student Sheet—Space Science Sequence 3.5

SOLAR SYSTEM TRAVEL BROCHURE GUIDELINES

Travel Brochure Checklist
Be sure to include the following information in your travel brochure:
- Five pieces of accurate and interesting information about your object in the brochure text. Write this information in a fun and exciting way—you are trying to convince someone that it would be fun to vacation there!
- Information about conditions on your object. For example:
  Does it have an atmosphere? If so, what is the atmosphere made of?
  What are its temperature conditions? Is it hotter or colder than Earth?
  What is the surface like? Is it rocky or gaseous?
  How big is it? Is it larger or smaller than Earth?
- At least one drawing of your object.
- A description of what it looks, smells, and feels like there. Remember, these descriptions should be based on accurate information about your object!
- A list of the research sources you used. In addition to the object fact sheet, you should use at least two other sources to gather information about your object. List these sources on the back of your brochure.
- Make your brochure exciting, colorful, neat, interesting, and factual! Be creative, funny, artistic, poetic, daring. The sky's the limit!
- Remember to write out all of the information in your brochure using your own words. Do not copy directly from your sources. Do not include anything you do not understand in your brochure.

Some Additional Ideas
- Explain what a person visiting should wear for their trip in order to be comfortable.
- Describe any special activities a person could do there.
- Come up with a cool nickname or slogan for your object.
- Make true-or-false-question cards that can be flipped over to reveal the correct answer underneath.
- Write a poem, rap, or song about your object. Have the words written out or the song recorded for visitors to appreciate when they visit your learning station!

Student Sheet—Space Science Sequence 3.5

---

12. **Assign Solar System object teams.** Read the list of student teams and their assigned Solar System objects. Distribute the Solar System Learning Station Guidelines sheet to each team.

13. **Distribute the Solar System Travel Brochure Guidelines to each student.**

14. **Go over the guidelines sheet with the class.** Have everyone fill in their assigned Solar System object, name, and project due date on the lines indicated. Go over each sheet's checklist with the class to make sure everyone understands the learning station assignment. Direct their attention to the travel brochure guidelines—they should include information about their object's temperature, atmosphere (if any), size, and composition. They should also use two other research sources in addition to their object fact sheet.

15. **Encourage students to be creative!** Tell students that they should be as creative as they can in designing their learning stations. Some fun ideas include: writing poems, making interesting illustrations, or building additional models. They can also come up with unique or strange activities (such as what it might feel like to jump on a planet with a different gravity than Earth) for their classmates to perform at the learning station.

16. **Remind students of what each learning station should look like.** Ask students to look at the illustration on their station guidelines sheet and point out that the station should include a title banner or sign, travel brochures, and an illustration and scale model of the object. Explain that each team will receive something in which to store their learning station objects.

Researching Assigned Objects, Creating Scale Models, and Designing Brochures

1. **Remind teams about the use of scale model materials.** Remind the class that teams with moons and small planets should use clay for their models, and that teams with large planets should use newspaper and masking tape. Teams working on the Asteroid Belt, Kuiper Belt, and Oort Cloud should use chalk dust or flour.

2. **Distribute team fact sheets.** Give the appropriate object fact sheets to each team—each student on a team should have his or her own copy. Point out where in the classroom you have placed the craft materials and additional research sources.

TEACHER CONSIDERATIONS

TEACHING NOTES

It's very important that students research their object's temperature, size, composition, and atmosphere. Later on, during their tour of the learning stations in Session 3.7, students will need to fill out object charts asking about these four characteristics.

To avoid too much classroom confusion and congestion, you could ask that one student from each team collect the art materials and research sources the team will need. Remind the class that they should share materials with one another.

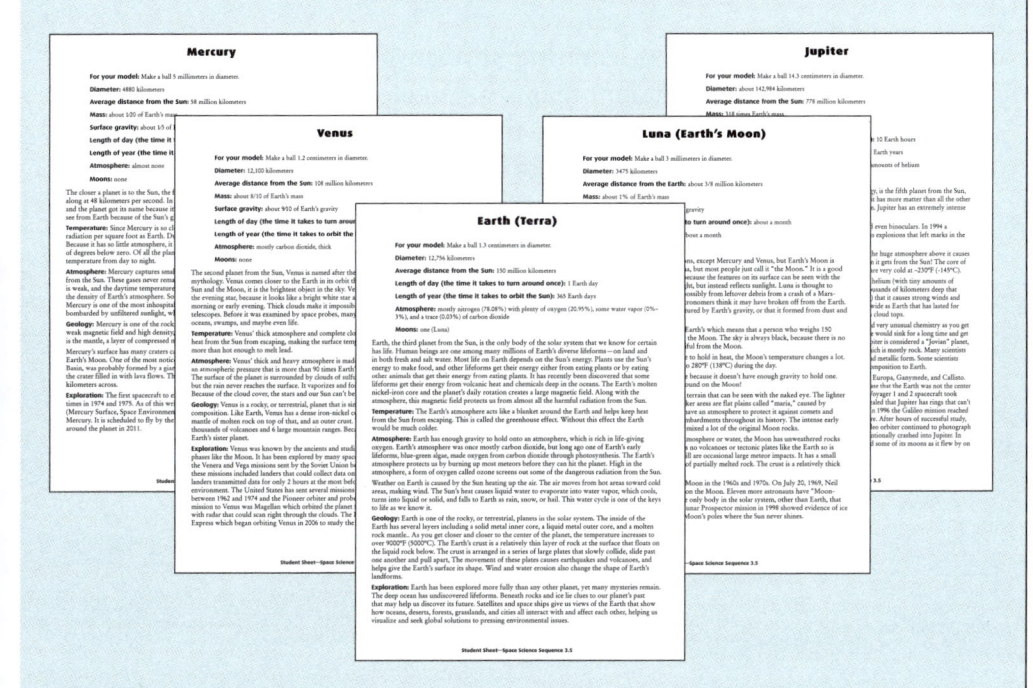

SESSION 3.5 Researching Objects in the Solar System

3. **Students begin work on learning station projects.** Have students begin work on their scale models, illustrations, and station banners. You may want to circulate around the classroom, offering suggestions or advice for any questions students may have.

4. **Students can begin designing their travel brochures after the other projects have been completed.** Make sure teams have finished making their scale models before allowing them to begin work on their brochures. If teams are taking longer to complete their scale models, signs, or color illustrations, assure them that there will be another session devoted entirely to working on the brochures.

5. **Cleaning up and storing learning station projects.** At the end of the class period, have student teams begin cleaning up. Give each team something in which to store their learning station creations. Ask them to put their object fact sheets in the storage container as well; they will need them again to continue work on their travel brochures. Make sure that students label the container with their names using a marker and masking tape.

TEACHER CONSIDERATIONS

TEACHING NOTES
Most teams should be able to complete their scale model by the end of the session. If for some reason this is not the case, give them a few minutes in the next session to finish their model. Another option is to assign completion of the scale model for homework.

SESSION 3.6
Completing Solar System Travel Brochures

Overview

In Session 3.5, students worked in pairs to complete projects for their assigned Solar System object's learning station. After completing a scale model and color illustration for their object, each student moved on to designing his or her own travel brochure. In this session, students continue work on their travel brochures, completing them in preparation for the next session's class tour of the learning stations. Some students may need additional time to complete their brochures—homework and extra class time are both options for extending this session. There are no key concepts for this session.

Completing Solar System Travel Brochures	Estimated Time
Finishing Travel Brochure Assignments	45 minutes
Total	**45 minutes**

What You Need

For the class:
- ❏ (optional) sample travel brochures
- ❏ a few pairs of scissors
- ❏ a few rulers (marked in centimeters)
- ❏ 16 glue sticks
- ❏ colored markers and/or colored pencils
- ❏ construction paper in different colors
- ❏ (optional) various research sources students can access
- ❏ assorted colors of modeling clay (optional: for students who need to complete their scale models)
- ❏ newspaper (optional: for students who need to complete their scale models)
- ❏ 1 roll of masking tape (optional: for students who need to complete their scale models)
- ❏ some chalk dust or flour (optional: for students who need to complete their scale models)

For each pair of students:
- ❏ additional copies of their assigned object's fact sheet from the copymaster packet or CD-ROM file

Getting Ready

Have all art materials and research sources organized and ready for use in an easily accessible location in the classroom, including additional copies of the fact sheets.

Unit Goals

The Solar System is centered around the Sun, the only star in the Solar System.

A wide variety of objects orbit the Sun in the Solar System.

Scientists categorize Solar System objects according to their characteristics; however, not all objects can be easily categorized.

Objects in the Solar System are in regular and predictable motion.

The Solar System is mostly empty space, and is very large compared to the objects located within it.

TEACHER CONSIDERATIONS

If students have not completed their scale models from Session 3.5, have them finish these first before allowing them to continue on to making their brochures!

Key Vocabulary

Scientific Inquiry Vocabulary

Category
Characteristic
Evidence
Model
Observation
Prediction
Scale
Scale model
Scientific explanation

Space Science Vocabulary

Asteroid
Astronomical Unit (AU)
Comet
Diameter
Heliosphere
Kuiper Belt Object (KBO)
Moon
Orbit
Planet
Sphere
Star
System

SESSION 3.6 Completing Solar System Travel Brochures

▶ Finishing Travel Brochure Assignments

1. **If necessary, re-read the Solar System object assignments to the class.** Do this only if students cannot remember to which object they have been assigned to create a travel brochure. (This will be highly unlikely!)

2. **Make sure students have their object fact sheets.** These should be stored in their learning station containers. Distribute fact sheets to students who have misplaced theirs.

3. **Point out where in the classroom you have placed the materials that students can use to create their brochures.**

4. **Students begin or continue work on brochures.** You may want to circulate around the classroom, offering suggestions or advice for any questions students may have.

5. **Cleaning up and putting away brochures.** At the end of the class period, have students put away their brochures in the same storage containers as their other learning station creations.

Many teachers have found that their students can easily spend several class periods working on their brochures. Decide whether they should finish their brochures today or whether you'll provide additional class time for them to complete their brochures, or have students finish them as homework.

TEACHER CONSIDERATIONS

TEACHING NOTES
To avoid too much classroom confusion and congestion, you may request that one student from each team collect the art materials and research sources the team will need. Remind the class that they should share materials with one another.

One teacher said, "I noticed that students were really starting to put facts together after a while. They were starting to see relationships because they had spent a lot of time discussing things while they were making the brochures."

SESSION 3.7
Taking a Tour of the Solar System

Overview

In the previous two sessions, students prepared learning stations for assigned Solar System objects by creating scale models and travel brochures for their object. They aimed to design unique learning stations that shared information about their objects in engaging or interesting ways. In this session, students "tour" the Solar System by visiting the learning stations made by their classmates. As students read one another's travel brochures, they learn interesting facts about the different objects. After completing their tour, students fill out object characteristic charts, which they then refer to in a class discussion about general planetary characteristic trends. During this session, the key concepts that will be added to the classroom concept wall are:

- *Solar System objects have a wide variety of characteristics.*
- *In general, the farther away a planet is from the Sun, the colder its temperature.*
- *The inner planets are smaller in size than the outer planets.*
- *The composition of the inner planets is mostly rocky, while the composition of the outer planets is mostly gaseous.*
- *With the exception of Mercury, all the planets have atmospheres.*

Taking a Tour of the Solar System	Estimated Time
Preparing for the Tour	5 minutes
Students Tour the Solar System	30 minutes
Discussing Planetary Characteristic Trends	10 minutes
Total	45 minutes

What You Need

For the class:
- ❑ overhead projector or computer with large-screen monitor or LCD projector
- ❑ prepared key concept sheets from the copymaster packet or CD-ROM file
- ❑ additional copies of Solar System object fact sheets from the copymaster packet CD-ROM file
- ❑ transparency of the Tour of the Solar System—Comparing Object Characteristics from the transparency packet or the CD-ROM file
- ❑ transparencies of the three pages of the Pre-Unit 3 Questionnaire (from Session 3.2) from the transparency packet or the CD-ROM file
- ❑ (optional) timer

For each pair of students:
- ❑ completed learning station projects (from Sessions 3.5 and 3.6)

Unit Goals

The Solar System is centered around the Sun, the only star in the Solar System.

A wide variety of objects orbit the Sun in the Solar System.

Scientists categorize Solar System objects according to their characteristics; however, not all objects can be easily categorized.

Objects in the Solar System are in regular and predictable motion.

The Solar System is mostly empty space, and is very large compared to the objects located within it.

TEACHER CONSIDERATIONS

The Tour of the Solar System—Travel Notes student sheet can be used to assess student understanding as an optional embedded assessment. See page 122 for more information.

Key Vocabulary

Scientific Inquiry Vocabulary
Category
Characteristic
Evidence
Model
Observation
Prediction
Scale
Scale model
Scientific explanation

Space Science Vocabulary
Asteroid
Astronomical Unit (AU)
Comet
Diameter
Heliosphere
Kuiper Belt Object (KBO)
Moon
Orbit
Planet
Sphere
Star
System

SESSION 3.7   Taking a Tour of the Solar System

You may want to have students approximate the general layout of the Solar System as they set up their stations. If so, start with Mercury and then position the other objects, in order, around the room.

Filling out the travel notes should not be a problem if each team of students has prepared their learning stations using their assigned object fact sheet. However, you may have a few teams who are less prepared or whose learning stations are lacking some information from the fact sheet.

For each student:
- ❑ completed travel brochure (from Sessions 3.5 and 3.6)
- ❑ 1 copy of the Tour of the Solar System—Travel Notes student sheet (4 pages) from the copymaster packet or CD-ROM file
- ❑ (optional) 1 copy of the Tour of the Solar System—Thought Questions student sheet from the copymaster packet or CD-ROM file

Getting Ready

1. **Prepare the key concept sheets.** Make a copy of each key concept and have them ready to post on the classroom concept wall during the session.

2. **Decide where each team should set up their learning stations.** You may want to take into consideration the flow of student traffic in the classroom.

3. **Make a copy of the Tour of the Solar System—Travel Notes student sheet (4 pages) for each student.**

4. **Make an overhead transparency of the Tour of the Solar System—Comparing Object Characteristics.** This will be used in a class discussion after students tour the stations.

5. **Optional: Make a copy of the Tour of the Solar System—Thought Questions student sheet for each student.** This student sheet contains several thought-provoking and fun questions for students to answer after they have completed their tours. Decide whether or not you would like your students to answer these questions. Another option would be to assign this as homework.

6. **Make additional copies of object fact sheets.** If necessary, an object fact sheet can be placed at a learning station if, during the tour, students feel that a particular station does not provide enough information about an object.

7. **Optional: Obtain a timer.** A timer that beeps or dings after two minutes can be useful for maintaining the pace of the tour. A non-distracting piece of "space-y" music about two minutes long can work as well.

TEACHER CONSIDERATIONS

If you had students research additional objects for learning stations, create a travel-note page with characteristic charts for these objects following the chart format shown on the student sheets. (An easy way to do this: make an extra copy of the student sheet, white-out the original object names on the charts, fill in any new object names, and then photocopy the modified charts for students to fill out.) Conversely, if you had to eliminate some objects from the tour, be sure to cross these objects out on the student sheets before copying them for students.

TEACHING NOTES

The key concepts can be posted in many different ways. If you don't want to use sentence sheets, here are some alternatives:

- Write the key concepts out on sentence strips.
- Write the key concepts out before class on a posted piece of butcher paper. Cover each concept with a strip of butcher paper and reveal each one as it is brought up in the class discussion.

SESSION 3.7 Taking a Tour of the Solar System

TOUR OF THE SOLAR SYSTEM—TRAVEL NOTES

Facts You Need to Know:
Earth's Size: Diameter is 12,800 km (7950 mi)
Earth's Temperature: Range is -88°C to 58°C (-127°F to 136°F)

MERCURY		VENUS	
Temperature		Temperature	
☐ hotter than Earth	☐ colder than Earth	☐ hotter than Earth	☐ colder than Earth
Size		Size	
☐ bigger than Earth	☐ smaller than Earth	☐ bigger than Earth	☐ smaller than Earth
What it is Made of		What it is Made of	
☐ mostly gas	☐ mostly rock	☐ mostly gas	☐ mostly rock
Atmosphere		Atmosphere	
☐ has an atmosphere	☐ has no atmosphere	☐ has an atmosphere	☐ has no atmosphere
Interesting Fact About Mercury:		Interesting Fact About Venus:	

EARTH'S MOON		MARS	
Temperature		Temperature	
☐ hotter than Earth	☐ colder than Earth	☐ hotter than Earth	☐ colder than Earth
Size		Size	
☐ bigger than Earth	☐ smaller than Earth	☐ bigger than Earth	☐ smaller than Earth
What it is Made of		What it is Made of	
☐ mostly gas	☐ mostly rock	☐ mostly gas	☐ mostly rock
Atmosphere		Atmosphere	
☐ has an atmosphere	☐ has no atmosphere	☐ has an atmosphere	☐ has no atmosphere
Interesting Fact About Earth's Moon:		Interesting Fact About Mars:	

If you opted for teams to use something other than the black trash bags in which to store their projects, you can still hand out the trash bags for students to use as a background for their station displays.

Emphasize to your students that the Tour of the Solar System—Travel Notes should take priority over the thought questions. Students will need to refer to their travel notes for the class discussion after the tour. The travel notes are also used to help illustrate this session's key concepts.

GO! Preparing for the Tour

1. **Give each student the Tour of the Solar System—Travel Notes student sheets to fill out.** Go over the chart format of the sheets with the class. Make sure everyone understands how to fill out the student sheets. Explain that, at each learning station, they should look for information about the four characteristics listed. Let them know whether you will be using their papers as an assessment tool. Say that if a learning station doesn't answer all of their questions, you will provide a fact sheet with more information for that object during the tour.

2. **Optional: If you've decided to have your students complete the Tour of the Solar System—Thought Questions student sheet as well, briefly go over the questions with them.** Tell students that the questions should be answered only **after** they have completed filling out their Tour of the Solar System—Travel Notes sheets.

3. **Have teams set up their stations at designated locations.** Circulate around the room, assisting students with preparations. If teams have stored their learning station projects in large, black, plastic trash bags, suggest that they use these bags as "space" backgrounds for displaying their work.

Students Tour the Solar System

1. **Preparing for take off!** Have all teams start off at the station they designed. Indicate in which direction they should travel when it's time for them to move on to the next station. Tell them that they will have about two minutes per station. You may choose to give them more time at each station, especially the first few they visit, if you have additional time or a longer class period.

2. **Students tour the Solar System by visiting learning stations.** Circulate through the classroom during the tour. If students indicate that a particular learning station doesn't provide enough information, leave a fact sheet at that station for visiting students to read. Let students know when it is time to move on to their next travel "destination."

3. **The tour concludes.** When all teams have finished visiting all of the stations, have everyone return to their seats. Tell them to use the information they've collected to respond to the questions on their Tour of the Solar System—Travel Notes student sheets.

462 • SPACE SCIENCE SEQUENCE 6–8   Session 3.7: Taking a Tour of the Solar System

TEACHER CONSIDERATIONS

TEACHING NOTES

The tour will probably take most of the class period. If it does, students can finish filling out their student sheets as homework or during additional class time. [Important Note: If you are unable to go through Discussing Planetary Characteristics after the tour, set aside time to debrief the class the next day. The discussion introduces this session's key concepts, and is important for helping students gain a better understanding of general planetary characteristic trends.] You may also go through the Discussing Planetary Characteristics discussion at the beginning of Session 3.8 before moving on to the outdoor scale model. (This is a good option if you will be teaching Session 3.8 the day after Session 3.7.)

ASSESSMENT OPPORTUNITY
EMBEDDED ASSESSMENT: TOUR OF THE SOLAR SYSTEM—TRAVEL NOTES

Student responses on the Tour of the Solar System—Travel Notes student sheet can be used as an embedded assessment. See the scoring guide on page 122 in the Assessment section.

One teacher suggested the following as an alternative to the class tour of the learning stations: Have teams give oral presentations about their objects to the class instead. A fun and creative twist on this option would be for teams to make "sales pitches" about their objects—their goal would be to convince others to come and visit their object. Remind everyone, though, that the information they present must be accurate and truthful.

SESSION 3.7 Taking a Tour of the Solar System

Tour of the Solar System—Comparing Object Characteristics

	Temperature (Earth's temperature: -127°F to 136°F)		Size (Earth's diameter: 12,800 km)		What it is Made of		Atmosphere	
	Hotter than Earth	Colder than Earth	Bigger than Earth	Smaller than Earth	Mostly gas	Mostly rock	Has an atmosphere	Has no atmosphere
Mercury								
Venus								
Earth's Moon								
Mars								
Asteroid Belt								
Jupiter								
Io								
Europa								
Callisto								
Saturn								
Titan								
Uranus								
Neptune								
Pluto								
Kuiper Belt								
Oort Cloud								

Discussing Planetary Characteristic Trends

1. **Solar System objects can be described using many characteristics.** Tell the class that, as they've probably just seen from their tour, Solar System objects can be described by a variety of characteristics. Ask students to share any interesting characteristics or facts they learned about an object during their tour. Post on the concept wall, under Key Space Science Concepts:

Solar System objects have a wide variety of characteristics.

2. **Explain the purpose of the Tour of the Solar System—Comparing Object Characteristics transparency.** Show students the transparency and point out that it will be used to record the same data as their travel notes, but in a different format. The travel notes are arranged by object, but the chart (on the overhead or computer monitor) is arranged to more easily make comparisons of the characteristics of those objects. Say that this reflects how scientists often look at the same data in different ways to understand it better.

3. **Looking for planetary characteristic trends.** Tell students that the class will begin discussing characteristics of the **planets only.** They will be looking for planetary characteristic trends. Let students know they should refer often to their completed travel notes during the discussion as the class works together to fill out the chart on the overhead. Point out to the class that the objects listed on the student sheet are in order, from closest to farthest from the Sun.

4. **Defining the terms *inner planet* and *outer planet*.** Ask students if they have ever heard of the terms *inner planet* or *outer planet*. Tell them that the planets closer to the Sun—Mercury, Venus, Earth, and Mars—are often called the inner planets. The planets farther away from the Sun—Jupiter, Saturn, Uranus, and Neptune—are called the outer planets.

TEACHER CONSIDERATIONS

TEACHING NOTES

Misconception Alert. If your students have studied the Earth's seasons (as in Unit 2 of this 6–8 Space Science Sequence), they should know that the small change in the Earth's distance from the Sun each year is not the reason for colder or hotter weather during different seasons. However, it's possible that students who previously held the misconception that the Earth's distance to the Sun causes seasons might re-adopt this incorrect idea. As students learn that there is a trend that relates the distance of different planets from the Sun to the average temperature of the planets, you may want to check for understanding that this is not the cause of Earth's seasons.

PROVIDING MORE EXPERIENCE

The Possibility of Finding Life in the Solar System. An optional topic for discussion during this session is whether or not any other objects in the Solar System might be capable of supporting life. This is a subject that greatly intrigues students, and if you have time to fit in this additional discussion, it is an ideal topic for students to assimilate their newly gained knowledge about various Solar System objects.

Discussing Conditions for Life

1. **Conditions suitable for life.** Ask students what conditions they think might make a planet or moon suitable for harboring life. [Students may mention things like air, water, moderate temperature, an atmosphere, sunlight, or gravity.] If necessary, assist students in their brainstorming by asking them to consider what conditions make Earth suitable for life. (Note: While this prompt may lead students to consider conditions that support life as we know it, it can be a good starting point for getting the discussion started.)

2. **Liquid water is necessary for life.** If no one mentions liquid water, tell students that most scientists think *liquid* water is a necessary condition for life. Explain that liquid water can only exist on a planet or moon if its temperature doesn't cause the water to solidify into ice or vaporize into gas. (On Earth, water freezes at 0°C [32°F] and boils at 100°C [212°F]. These temperatures are not necessarily the same on other planets or moons, however, since they will vary according to the object's atmospheric pressure.)

continued on page 467

SESSION 3.7 Taking a Tour of the Solar System

5. **The temperature of planets.** Now ask students to look at their travel notes. Refer them to the boxes they checked under Temperature for each planet. Beginning with Mercury, ask students to report on the temperatures of the planets as you fill out the projected transparency. After you have completed the temperature columns, ask students if they notice a temperature trend among the planets. [Planets beyond Earth are colder than Earth. Venus is hotter than Earth. Mercury can be colder or hotter than Earth, depending on which side of Mercury you are considering.] Tell the class that the temperature of planets follows a general trend. Post on the concept wall, under Key Space Science Concepts:

   In general, the farther away a planet is from the Sun, the colder its temperature.

   Be sure to emphasize that Mercury is an exception to the rule—while the side facing the Sun is extremely hot, the side facing away from the Sun is extremely cold—as cold as the *maximum* temperature on Pluto! Also point out that Venus is slightly unusual, too. Its thick carbon dioxide atmosphere acts like a greenhouse, trapping heat. Venus gets hotter than Mercury's hot side!

6. **The size of planets.** Now direct the class to look at the boxes they checked under Size for each planet. Again, beginning with Mercury, ask students to report on the size of the planets as you fill out the transparency. After you have completed the size columns, ask students if they notice a size trend among the planets. [The outer planets are bigger than the inner planets. The outer planets are bigger than Earth. The inner planets are smaller than Earth.] Post on the concept wall, under Key Space Science Concepts:

   The inner planets are smaller in size than the outer planets.

7. **The composition of planets.** Next, ask students to look at the boxes they checked under What it is Made Of for each planet. Ask students to report on the composition of the planets as you fill out the transparency. After you have completed the What it is Made Of columns, ask students if they can identify a composition trend for the planets. [The inner planets are made mostly of rock. The outer planets are made mostly of gas.] Tell the class that this is why the inner planets are sometimes also called the terrestrial planets and why the outer planets are sometimes called the gas giants. Post on the concept wall, under Key Space Science Concepts:

   The composition of the inner planets is mostly rocky, while the composition of the outer planets is mostly gaseous.

TEACHER CONSIDERATIONS

PROVIDING MORE EXPERIENCE
continued from page 465

3. Why is liquid water a key factor? About two-thirds of the weight of cells in living organisms is accounted for by water, which gives cells many of their properties. Living organisms use water as a solvent in which nutrients can be dissolved and transported to all parts of the organism. This is true for both plants and animals. The blood in animals (which is mostly water) also transports oxygen to all parts of the organism. Water can only serve this kind of purpose if it is in liquid form.

4. Have students suggest objects that may be able to support life. Encourage students to refer to their Tour of the Solar System—Travel Notes to come up with candidates. Remind students that liquid water, a vital component for life, can only exist on a planet or moon if its temperature doesn't cause the water to solidify into ice or vaporize into gas. Ask, "Where do you think it's most likely for us to find life in the Solar System besides Earth?" "Why?"

Discussing candidates. Students may choose Mars, Europa, or Titan as possible candidates for life. They may name another planet or moon. Or they may say that no other places in the Solar System seem habitable. In any case, they should cite the characteristics of the places they mention and use the characteristics to back up their argument. In general:

- A place with a rocky surface might be more likely to harbor life.
- Liquid water is often considered a prerequisite for the development of life as we know it.
- Places with moderate temperatures are more likely to harbor life. (Students should know that the outer Solar System is generally too cold to support life as we know it. However, some moons are heated by volcanic activity. Europa may have liquid water under its ice.)
- Places with atmospheres are more likely to harbor life. (Atmospheres protect against harmful energies from the Sun, keep temperatures more stable, help living things breathe, etc.)

Misconception Alert: A narrow definition of the term *life*. You may want to remind students that references to conditions for "life" on other planets or elsewhere do not necessarily mean "intelligent life," or life as we know it. Some students my assume that only humans are being discussed but, in fact, "life" can mean microorganisms or other simple forms that do not exist on Earth that are alive and have adapted to extreme conditions.

continued on page 469

SESSION 3.7 Taking a Tour of the Solar System

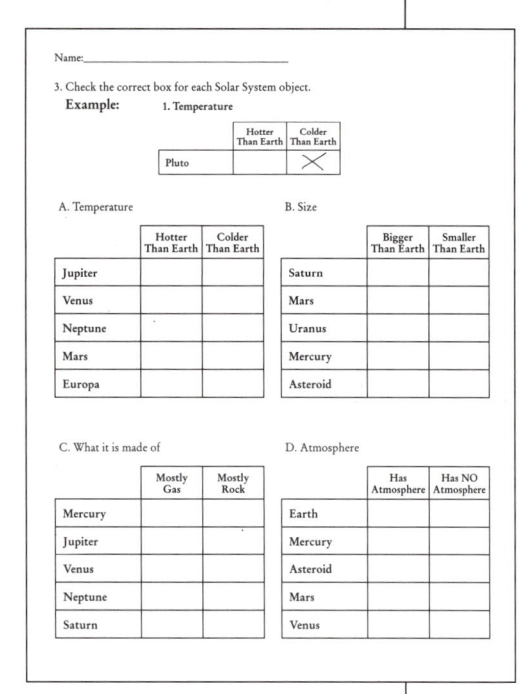

8. Planets with and without atmospheres. Tell students to look at the boxes they have checked under Atmosphere for each planet. Once more, ask students to report on the atmosphere of the planets as you complete the transparency. Then ask which planets have atmospheres and which do not. [All of the planets have atmospheres except for Mercury.] Post on the concept wall, under Key Space Science Concepts:

With the exception of Mercury, all the planets have atmospheres.

9. Revisit Question #3 on the Pre-Unit 3 Questionnaire with students. Place the transparency of the Pre-Unit 3 Questionnaire on the overhead projector or project it on the large-screen monitor. Cover up all of the questions on the questionnaire except for Question #3. Go through the four characteristics charts with the class and ask students which boxes they would check now that they have learned more about the characteristics of different Solar System objects.

10. Clean up learning stations and collect scale models. Have student teams clean up and put away their learning stations. Be sure to collect all of the scale models from each station. Tell the class that their scale models will be used in the next session.

TEACHER CONSIDERATIONS

PROVIDING MORE EXPERIENCE
continued from page 467

Misconception Alert: students may assume that we have already found life on other planets. If you try asking a class full of students whether they have heard that scientists have detected life on other bodies in the Solar System, it is quite likely that some students will say yes. They may have come across some tabloid articles that were made up for the entertainment of the reader. Or perhaps they have heard one of the legitimate reports about scientists having discovered or theorized about one of the preconditions for life on another planet or moon—not life, and not even all the conditions for life, but usually just one possible condition for life. There have been a few claims based on presumed astronomical evidence for extraterrestrial life. At the beginning of the 20th century, several astronomers believed they had observed canal-like structures on Mars that, in their view, could only have been created by intelligent living creatures. Bigger and better telescopes later revealed that these "canals" were either naturally occurring canyons or simply did not exist.

Another claim for the possibility of life on Mars was based on a meteorite found in Antarctica in 1984. Ten years later, scientists agreed that evidence showed that the meteorite did indeed come from Mars. The "big news" was that the meteorite had tiny structures in it that resembled fossilized bacteria. There were some chemical indications that these structures could have originated from living organisms, but over time, scientists have come up with alternative explanations, which seem more plausible. Scientists—including those who originally put forward the idea—no longer consider this meteorite proof that life ever existed on Mars.

Most current ideas about possible life in the Solar System are very tentative, partial, and preliminary. Astronomers generally agree that they (and the world!) will need to wait for much more convincing evidence before claiming that there is life "out there." However, the possibilities are intriguing and the search very exciting and compelling. One such instance is Europa, a moon of Jupiter. The evidence is strong that the surface of Europa is a shifting layer of ice, which conceals an ocean of liquid water. Liquid water is considered one of the essential ingredients for the development of life. Colors in the ice of Europa remind scientists of ice on Earth, which is colored by bacteria. However, that, in itself, is not evidence that bacteria exist on Europa. This is an item of great interest to scientists and all those intrigued by the search for extraterrestrial life, but it does not come close to a convincing suggestion that there actually is life on Europa.

SESSION 3.8

Outdoor Scale Model of the Inner Solar System

Overview

Most models of the Solar System emphasize the order and physical appearances of the planets. However, very often the sizes of the planets and the distances separating them from one another are not presented **to scale**. For this reason, many students are unaware that empty space—not the collection of planets—is the largest component of the Solar System. In this session, students begin by organizing the planet scale models collected from their learning stations. As they order each planet according to its distance from the Sun, they observe that the planets' orbits naturally divide the Solar System into two regions. The inner Solar System consists of small, rocky planets, while the outer Solar System consists of large, gaseous planets. The class then heads outside to construct an outdoor scale model of the inner Solar System. As students pace out and visualize the orbits of the four inner planets, they begin to realize that the Solar System consists primarily of the empty space between the planets. The class considers the orbits of the outer planets and completes the Solar System scale model in Session 3.9. There are no key concepts for this session.

Outdoor Scale Model of the Inner Solar System	Estimated Time
Reviewing Characteristics of the Planets	5 minutes
Organizing the Sun and Planets	10 minutes
Introducing the Scale Model	5 minutes
Walking Through the Outdoor Scale Model	25 minutes
Total	**45 minutes**

What You Need

For the class:
- ❑ a paper circle about 139 centimeters in diameter (for a scale model representation of the Sun)
- ❑ scale models of all of the planets (from the learning stations in Session 3.7)
- ❑ tape
- ❑ meter stick

Getting Ready

1. **Consider arranging for an adult volunteer to assist you with this session.** You may find it helpful to have an adult volunteer assist you with class management for the outdoor scale model. Consider asking a fellow faculty member, a teacher's aide, or a parent to join the class for this session. Most adults are impressed by the scale model activity the first time they see it done.

Unit Goals

The Solar System is centered around the Sun, the only star in the Solar System.

A wide variety of objects orbit the Sun in the Solar System.

Scientists categorize Solar System objects according to their characteristics; however, not all objects can be easily categorized.

Objects in the Solar System are in regular and predictable motion.

The Solar System is mostly empty space, and is very large compared to the objects located within it.

TEACHER CONSIDERATIONS

TEACHING NOTES
The preparation for this session is more extensive than usual. In addition to assembling activity materials, you will need to find an appropriate outdoor space for the scale model activity. Prepare for this session at least a day before you plan to run the model with your students. You may also want to have a backup activity planned in case unexpected bad weather forces you to postpone the outdoor scale model activity.

ASSESSMENT OPPORTUNITY
CRITICAL JUNCTURE: UNDERSTANDING MODELS AND SCALE MODELS
Before doing the scale model activities in this session and Session 3.9, check that your students have a good understanding of why scientists use models and what a scale model is. If needed, review these ideas with your students using the discussion of models outlined in Session 3.1.

Key Vocabulary

Scientific Inquiry Vocabulary
Category
Characteristic
Evidence
Model
Observation
Prediction
Scale
Scale model
Scientific explanation

Space Science Vocabulary
Asteroid
Astronomical Unit (AU)
Comet
Diameter
Heliosphere
Kuiper Belt Object (KBO)
Moon
Orbit
Planet
Sphere
Star
System

SESSION 3.8 Outdoor Scale Model of the Inner Solar System

A space that is 150 meters long will allow the class to walk out to the scaled distance of Earth's orbit. A space that is 228 meters long will allow the class to walk out to the scaled distance of Mars' orbit.

2. **Prepare a paper circle with a diameter of 139 cm as a scale model of the Sun.** Cut the circle out from a large sheet of butcher paper or several pieces of paper taped together. (If possible, use yellow construction paper.)

3. **Gather all the scale models of the planets.** You will need the scale models of the planets that the students made for the learning stations and the scale model of Earth (13 mm ball of clay) that you made for Session 3.5.

4. **Find a large outdoor space in your school where the class can pace out the scale model.** You will need a space that is at least 150 meters long, but 228 meters is preferable. Find a wall or fence at one end of the space where you can tape up the paper scale model of the Sun. This will be the starting point for the scale model.

5. **Practice pacing out scale model distances for the inner planets.** One meter is approximately one *large* step. Use the meter stick to help you determine how large your steps should be. Don't worry about measuring distances exactly for this model; approximate distances are sufficient to illustrate the point of this activity to the class. You're just pacing this off now to be sure you'll have enough space and to take note of the approximate location of each planet.

   a. **Start at the end of the space where you intend to tape up the Sun.**

   b. **Walk to Mercury.** Pace out 58 meters from the Sun.

   c. **Walk to Venus.** Pace out 50 more meters from Mercury. (You should be 108 meters away from the Sun.)

   d. **Walk to Earth.** Pace out 42 more meters from Venus. (You should be 150 meters away from the Sun.)

   e. **Walk to Mars.** Pace out 78 more meters from Earth. (You should be 228 meters away from the Sun.)

   If you do not have space to include Mars in the model, identify a landmark (such as a tree or a building) at nearly the correct distance. Point it out to your students at the end of the outdoor scale model activity so they can visualize the location of Mars' orbit.

TEACHER CONSIDERATIONS

TEACHING NOTES

An easy way to measure out the correct diameter for the circle.
Make a compass using a strip of cardboard that is slightly longer than 69.5 cm (27⅜ inches), which is the radius of the model Sun. Make a hole at each end of the strip, 69.5 cm apart. Put a pencil point or a push-pin in one hole and hold it fixed in the center of the paper. Put a pencil point in the other hole and move it around the center to draw the circle. This can also be done with a string tied to a pencil, although it is more difficult to keep the length accurate. (Note: It's not important for the measurement to be completely precise. If the diameter of the circle is within a few centimeters of the desired 139 cm, the circle will be sized well enough for the purposes of the scale model.) To save time, you may want to ask a student or an adult volunteer to make the scale model of the Sun for you.

One giant step makes an easy approximation of a meter, but some people may be more comfortable taking two smaller steps for each meter. Practice with a meter stick.

SESSION 3.8   Outdoor Scale Model of the Inner Solar System

GO! Reviewing Characteristics of the Planets

1. **Discussing planetary characteristics.** Tell the class that last session's tour helped them learn about the characteristics of many different objects in the Solar System. Remind them that after the tour, the class discussed some general characteristics of the planets.

2. **Reviewing characteristics of the inner planets.** Ask students which planets are considered the inner planets. [Mercury, Venus, Earth, and Mars.] Ask what characteristics define the inner planets. [They are small and mostly rocky.]

3. **Reviewing characteristics of outer planets.** Ask students which planets are considered the outer planets. [Jupiter, Saturn, Uranus, and Neptune.] Ask what characteristics define the outer planets. [They are large and gaseous.] If students ask about Pluto, note that the characteristic of being large and gaseous is one of Pluto's characteristics that does not match the outer planets. They should be on the lookout for more as lessons continue.

Organizing the Sun and Planets

1. **Place the scale models of all the planets on a table or shelf in the classroom.** Do not place the planets in their correct order. Make sure that the models are visible to everyone in the class. The various moons, Asteroid Belt, Kuiper Belt, and Oort Cloud are hard to incorporate into this scale model, so they should not be included.

2. **Display the paper scale model of the Sun.** Tape it to a wall or chalkboard near the models of the planets. Tell students that this flat, two-dimensional model of the Sun was made using the same scale as their learning station planet models. Students may express surprise at how large the model of the Sun is. (**Make sure they understand that, in reality, the Sun is spherical and three-dimensional**.)

3. **Put the planets in order.** Ask the class to help you put the planets in their correct order. Starting from the Sun, ask students, "Which planet comes next?" until all of the models are ordered according to their distance from the Sun. If any students point out that the distances between the planets are not being represented to scale, tell them that that would be an excellent way to create a good scale model. Tell the class that they will be doing that shortly in the next activity.

If necessary, remind students that a scale model is a representation of an object that is smaller or larger than the actual size of the real object. Every part of the real object is measured and made smaller or larger by the same factor to create the scale model.

TEACHER CONSIDERATIONS

TEACHING NOTES
If you did not have a chance to go through the Discussing Planetary Characteristics Trends discussion at the end of Session 3.7, use the review at the beginning of this session to debrief the class—go through the more extensive discussion outlined in Session 3.7 on pages 464–468. (Note that the discussion from Session 3.7 requires 10 minutes to complete.)

SESSION 3.8   Outdoor Scale Model of the Inner Solar System

4. **Identifying the inner and outer planets.** With the planets lined up and in order, ask the class if they notice anything interesting about the planets and how they are arranged in the Solar System. Students may observe that:

   a. **The first four planets are smaller.** Remind the class that as just discussed, these planets are often called the "terrestrial planets" or "inner planets." These mostly rocky bodies comprise the inner Solar System.

   b. **The next four planets are significantly larger.** Remind the class that as just discussed, these planets are often called the "gas giants" or "outer planets." These mostly gaseous bodies comprise the outer Solar System.

5. **Remind the class of the key concepts from the previous session.** To reinforce what the class has just discussed and observed, refer to the following posted key concepts from Session 3.7:

   The inner planets are smaller in size than the outer planets.

   The composition of the inner planets is mostly rocky, while the composition of the outer planets is mostly gaseous.

Introducing the Scale Model

1. **Introduce the activity.** Tell the class that today they will be using the models of the planets to create a scale model of the inner Solar System. This model will show them not only the scaled *sizes* of the four inner planets, but the scaled *distances* between them as well.

2. **In this model, distance and size are both represented by the same scale.** Tell students that the planet models are scaled to one-billionth the size of the real planets they represent. Explain that in this model, the same scale will be applied to the distances separating the planets from one another. This means that the distances shown in this model will be one-billionth what they are in reality.

3. **In this model, one meter represents one million kilometers.** Show the class the meter stick and tell them (or have them calculate) that on a scale of one to one billion, a meter will represent a distance of one million kilometers.

TEACHER CONSIDERATIONS

TEACHING NOTES

Another way to explain the scale. Each model is one-billionth the size of the object it represents. In other words, the real object is one billion times larger than the size of the scale model.

To explain the scale calculation to your students: A scale of one to one billion means that distances in this model will be one-billionth of their size in reality. That means that one meter represents one billion meters. Since one billion meters equals one million kilometers, in this model one meter represents one million kilometers.

SESSION 3.8 Outdoor Scale Model of the Inner Solar System

4. Figuring out scaled distances of the planets from the Sun. Standing by the scale models of the Sun and the planets, tell the class that Mercury is about 58 million kilometers away from the Sun. Ask, "In the scale model we are about to create, how far away should Mercury be from the Sun?" [58 meters would be the scaled distance of 58 million kilometers.] Next, tell the class that Venus is about 108 million kilometers away from the Sun and ask what Venus' scaled distance should be [108 meters]. Repeat this with Earth and Mars as well. [Earth is about 150 million kilometers away, and Mars is about 228 million kilometers away.] Tell everyone that the class will now move outside to model the distances of the inner planets from the Sun.

Walking Through the Outdoor Scale Model

1. Take the class outside to begin assembling the scale model. Ask four students to carry the scale models of Mercury, Venus, Earth, and Mars. Take the paper scale model of the Sun, a roll of tape, and the meter stick. Lead the class to the space you chose earlier and tape the scale model of the Sun up on a wall or fence at one end of the space.

2. Show your students what a meter-long step looks like. Invite them to try a similar step themselves. Tell students that using the scale discussed in class earlier (where one meter equals one million kilometers), each giant meter-long step will carry them about one million kilometers across the Solar System.

3. Walk from the Sun to Mercury's orbit. Remind the class that the scaled distance from the Sun to Mercury is 58 meters. Have the class walk with you 58 meters away from the Sun. Tell students that they are now at the distance of Mercury's orbit. Ask the student carrying the model of Mercury to hold it up for everyone to see. Ask how Mercury's size compares to its distance from the Sun. [Mercury's size is tiny in comparison to its great distance from the Sun.]

4. Visualize the complete orbit of Mercury. Ask someone to describe how Mercury moves around the Sun. [It moves in an almost circular orbit, staying at about the same distance away from the Sun.] Give students a minute to imagine Mercury's orbit at this scale. Have students point out objects that Mercury would pass near on its orbit around the Sun. Leave a pair of students behind with the model of Mercury.

TEACHER CONSIDERATIONS

SESSION 3.8 Outdoor Scale Model of the Inner Solar System

5. **Walk from Mercury's orbit to Venus' orbit.** Tell the class that since the scaled distance from the Sun to Venus is 108 meters, they now need to pace out 50 more meters to reach the distance of Venus' orbit. Walk out 50 meters and ask the student carrying the model of Venus to hold it up for everyone to see. Call back to the student pair at Mercury and ask them to describe the appearance of Venus from Mercury. Have the students with you at Venus respond by describing what Mercury looks like from Venus.

6. **Visualize the complete orbit of Venus.** As with Mercury, have students describe the path that the model of Venus would take in its nearly circular orbit around the model Sun. Note that at the place you are standing now, Mercury and Venus are as close together as they can be. When they are at other points in their orbits, they can be much farther apart. Leave a pair of students behind with the model of Venus.

7. **Walk from Venus' orbit to Earth's orbit.** Tell the class that Earth's scaled distance from the Sun is 150 meters. This means the class needs to pace out 42 more meters (for a total of 150 meters from the Sun) to the distance of Earth's orbit. Walk out 42 meters and ask the student carrying the model of Earth to hold it up for everyone to see. Call to the student pairs at Mercury and Venus and ask them to describe how Earth appears to them. Then have the students at Earth respond with a description of how the Sun, Mercury, and Venus appear to them from the distance of Earth's orbit.

8. **Have the Mercury and Venus student pairs rejoin the class.** Call to the student pairs and have them join the class at Earth's orbit. Ask them to bring the models of Mercury and Venus with them.

9. **Define the term *Astronomical Unit*.** Tell the class that astronomers use a specific term to describe the great distance from the Sun to the Earth. This distance is called an *Astronomical Unit*, or AU for short. Explain that the term AU can make it much easier to describe and compare really large distances in the Solar System. For example, it is more convenient to say that the Earth is one AU from the Sun than it is to say that it is 149.6 million kilometers away from the Sun.

TEACHER CONSIDERATIONS

SESSION 3.8 Outdoor Scale Model of the Inner Solar System

10. **Tell the class that Mars' orbit is about 1.5 AU from the Sun.** With this information, have students estimate where Mars' orbit would be from the distance of Earth's orbit. If your space is large enough for the class to walk out to Mars, pace out an additional 78 meters (for a total of 228 meters from the Sun) to the distance of Mars' orbit. If your space isn't large enough for the class to walk out to Mars, point out an appropriately distanced landmark and walk with your students as far toward it as possible. Have the class look back toward the Sun and see how small it looks. Explain that although the real Sun would be a billion times bigger than it currently appears, it would also be a billion times farther away than it currently is! The scale model Sun that the class is looking at is exactly the same size that the real Sun would *appear to be* from the real Mars!

11. **Ask students to describe the space between the inner planets.** Ask the class how they would describe the spacing of the inner planets. Ask, "Are the planets located very closely to one another?" [No, the inner planets are not too close to one another—there is a lot of space between them.] Ask the students what seems to make up most of the inner Solar System. [Empty space!]

12. **Including the outer planets next time.** Tell the class that today's scale model can only be done for the four inner planets of the Solar System. Ask the class why they think this is. [Distances to the outer planets are even greater. It would take too long to walk out to the orbits of the outer planets.] Explain that in order to include the outer planets in this scale model, the class will need to finish constructing the model on a map in the classroom during the next session.

13. **Walk back to the classroom.** Bring all materials back with you. Be sure to pick up the paper scale model of the Sun as you return to class.

TEACHER CONSIDERATIONS

One teacher said, "The students enjoyed predicting where the planets would be and how many we could place on the football field. We used a metric measuring wheel to pace off our meters. That worked really well. When we returned to class, I could see their brains working! It was a revelation to discover that Pluto's orbit was almost out to the shopping mall. This was an exciting way to make an abstract concept so concrete for my kids. They got it!"

SESSION 3.9

The Outer Solar System and Beyond

Overview

In the previous session, students observed that planets in the inner Solar System shared characteristics that were different than those shared by planets in the outer Solar System. They then participated in a walking outdoor scale model of the inner Solar System, pacing as a class out to the distances of Mercury, Venus, Earth, and Mars. By walking out to the orbits of the inner planets, students visualized the great distances between these objects. In this session, students finish the scale model by looking at the orbits of the outer planets drawn to scale on a map of the school. After discussing the large distances between the outer planets, students contemplate the question, "Where does the Solar System end?" by reading about the NASA Interstellar Boundary Explorer (IBEX) mission. Then, student groups discuss how an organized system like the Solar System might have formed. Using a simple model, students observe how a disorderly collection of matter might have evolved into a more orderly system of orbiting objects. During this session, the key concepts that will be added to the classroom concept wall are:

- *The Solar System is mostly empty space and is very large compared to the objects within it.*
- *The Solar System formed from gas and dust, which gravity drew together into a whirling system.*

The Outer Solar System and Beyond	Estimated Time
Completing the Scale Model on a Map	10 minutes
Discussing the Scale Model	10 minutes
Where Does the Solar System End?—Student Reading	15 minutes
A Swirling Cloud Model	10 minutes
Total	**45 minutes**

What You Need

For the class:
- ❏ overhead projector or computer with large-screen monitor or LCD projector
- ❏ prepared key concept sheets from the copymaster packet or CD-ROM file
- ❏ prepared transparency of a local map that includes the school and extends at least five kilometers from the school (see Getting Ready)
- ❏ transparencies of the three pages of the Pre-Unit 3 Questionnaire (from Session 3.2) from the transparency packet or the CD-ROM file
- ❏ scale models of all the planets (from the learning stations in Session 3.7)

Unit Goals

The Solar System is centered around the Sun, the only star in the Solar System.

A wide variety of objects orbit the Sun in the Solar System.

Scientists categorize Solar System objects according to their characteristics; however, not all objects can be easily categorized.

Objects in the Solar System are in regular and predictable motion.

The Solar System is mostly empty space, and is very large compared to the objects located within it.

TEACHER CONSIDERATIONS

Key Vocabulary

Scientific Inquiry Vocabulary

Category
Characteristic
Evidence
Model
Observation
Prediction
Scale
Scale model
Scientific explanation

Space Science Vocabulary

Asteroid
Astronomical Unit (AU)
Comet
Diameter
Heliosphere
Kuiper Belt Object (KBO)
Moon
Orbit
Planet
Sphere
Star
System

SESSION 3.9 The Outer Solar System and Beyond

- ❏ compass (for drawing circles on the map)
- ❏ black pepper in a shaker
- ❏ paper towels or a few rags (in case of spills)

For each team of 4-6 students:
- ❏ 1 flat-bottomed, round container made of clear or light-colored material
- ❏ 1 stirring implement (a stir stick, a popsicle stick, or a spoon)

Cottage cheese containers or wide-bottomed plastic cups will also work.

For each student:
- ❏ 1 copy of Where Does the Solar System End?—Student Reading from the copymaster packet or the CD-ROM file
- ❏ 1 copy of Where Does the Solar System End?—Reading Questions from the copymaster packet or the CD-ROM file (optional: see the Providing More Experience note on page 495.)

Getting Ready

1. Prepare the key concept sheets. Make a copy of each key concept and have them ready to post onto the classroom concept wall during the session.

2. Find a local map that includes your school and surrounding neighborhood. Check that the scale of the map represents a kilometer as 2–4 centimeters. Many common local road maps will work, or you can print a map from an Internet site such as MapQuest (www.mapquest.com) or Google Maps (maps.google.com). The Internet option has the advantage that you can choose the scale and can position your school so the orbits of the outer planets will fit.

3. Draw scaled orbits of the outer Solar System planets onto the map.

   a. Locate your school. Mark the approximate spot on the map where you taped up the paper scale model of the Sun in Session 3.8's outdoor scale model.

   b. Draw a circle around the Sun to represent the orbit of Mars. Use the scale of the map to set the radius of Mars' orbit at just under 0.25 or ¼ kilometers. Using a compass, draw a circle around the Sun on the map to indicate Mars' orbit. (The circle should only be roughly 1–2 centimeters in diameter, depending on the scale of the map.)

   c. Draw in orbits for the rest of the outer planets. As with Mars, use your map's scale to set your compass to the correct scaled radius for each planet. Draw in the orbits of Jupiter, Saturn, Uranus, and Neptune. For any orbits that are too large to fit completely on the map, draw the part of the circle that does fit.

TEACHER CONSIDERATIONS

TEACHING NOTES
The key concepts can be posted in many different ways. If you don't want to use sentence sheets, here are some alternatives:

- Write the key concepts out on sentence strips.
- Write the key concepts out before class on a posted piece of butcher paper. Cover each concept with a strip of butcher paper and reveal each one as it is brought up in the class discussion.

TEACHING NOTES
If you can't find a map with the correct scale, use a photocopier to make a copy of the map with an increased or decreased scale.

SESSION 3.9 The Outer Solar System and Beyond

d. Optional: Draw in the Asteroid Belt. If you choose to include it on your map, lightly draw a circle with a radius representing 0.3 km on the map and another with a radius representing 0.5 km. Color in the region between the circles to represent the region of the Asteroid Belt that contains over 90% of the asteroids.

4. Photocopy a section of the map onto an 8½" x 11" transparency. Copy a section of the map that includes the school and a region that extends at least five kilometers from the school. (This will ensure that a section of Neptune's orbit is included on the transparency.) You may have to make the transparency with your school situated at one end in order to include all of the outer planets' orbits. When deciding which part of the map to copy, choose an area with landmarks that may be familiar to your students.

5. Copy Where Does the Solar System End?—Student Reading. Each student should have his or her own copy.

6. Decide how you will divide the class into teams of 4–6 students for the A Swirling Cloud Model activity.

7. Prepare each team's container for the swirling cloud model. Fill each container with about 1½" of water. Set the containers aside for distribution later. Have a stirring implement ready for each container and the pepper shaker handy.

8. Familiarize yourself with the model. Use one of the containers to try out A Swirling Cloud Model activity as described on page 496. (This takes only a minute.) Replace the peppered water with clean water when you are finished.

Be sure to save the prepared overhead transparency of the map. You will need it for Session 3.10.

TEACHER CONSIDERATIONS

TEACHING NOTES

If your compass is not wide enough for the larger orbits, design a compass out of a strip of cardboard. For a more detailed description about using a strip of cardboard as a compass, refer to the example described in Session 3.8 on page 473.

SESSION 3.9 The Outer Solar System and Beyond

GO! Completing the Scale Model on a Map

1. **Briefly review the outdoor scale model.** Ask some students to share what they remember about last session's outdoor scale model of the inner Solar System. Say that today they will complete the model by including planets in the outer Solar System. The class will do this by looking at the orbits of the outer planets drawn on a map of the school and its surrounding neighborhood. Emphasize that today's scale model will use the same scale as the one the class used for the outdoor model. The only difference is that the class will not pace out the scaled orbit distances—these distances will be represented on a map of the neighborhood around the school.

2. **Show the prepared overhead transparency of the map with the orbits of the outer planets drawn in.** Have students locate the school on the map. Ask them to identify the approximate location of the Sun from last session's outdoor model.

3. **Point out the circle that represents the orbit of Mars.** Remind the class that in the outdoor scale model, they were able to walk out to the distance of Mars' orbit. The orbits of the other inner planets (Mercury, Venus, and Earth) are all within the orbit of Mars but are too small to show easily on the map. Tell the class that this mapped scale model will help them visualize the orbits of the outer Solar System planets.

4. **Move outward from Mars through all the orbits of the planets.** For each planet, have the class identify a familiar landmark on the map at about the distance of the planet's orbit. For example, a student might point out: "Mars' orbit could be at the gas station on Main Street." Hold up the scale model of Mars and ask the class to imagine how small Mars would appear if they were to look at it from the school. Have them imagine how long it would take for them to walk out to Mars' landmark. Repeat this for all the outer planets.

5. **Even larger distances to consider.** After going through all of the outer planets, tell the class that the Kuiper Belt and Oort Cloud are located at even *greater* distances from the Sun than Neptune's orbit. Using the same scale as before, the Kuiper Belt would be 4.5–8 kilometers away, and the Oort Cloud would extend out to about 9,000 kilometers away! Proxima Centauri, the nearest star to our Sun, would be at a mind-boggling distance of 41,000 kilometers away.

TEACHER CONSIDERATIONS

SESSION 3.9 The Outer Solar System and Beyond

Discussing the Scale Model

1. **One scale model formed from two separate models.** Tell the class that today they completed the scale model of the Solar System they started last time when they paced out the orbits of the inner planets. Since the orbits of the outer planets were much too large to pace out, the class used a scale model of the outer planets drawn on a map of the school and its surrounding neighborhood.

2. **The spacing of the orbits of the inner planets.** Using the overhead transparency, point out to the class how crowded the orbits of the inner planets would appear to be if they had been drawn in. Ask the class to think back to when they paced out the orbits of the inner planets. Ask, "Were the orbits really spaced closely together?" "Did the orbits seem crowded in the outdoor scale model?" [No, there was a lot of empty space between each orbit.]

3. **The spacing of the orbits of the outer planets.** Now ask the class to consider the orbits of the outer planets on the transparency. Point out that these orbits are spaced even farther apart than the orbits of the inner planets. Tell the class that if they were to make an outdoor scale model with the outer planets included, they would have to pace out considerably larger distances to get to each planet's orbit!

4. **The Solar System consists mostly of empty space.** Ask the class what the Solar System seems to have a lot of. [Empty space.] Point out that all the objects in the Solar System are very small in comparison to the size of the Solar System itself. Post on the concept wall, under Key Space Science Concepts:

   The Solar System is mostly empty space and is very large compared to the objects within it.

5. **Discuss the plane of orbit of the planets.** Remind students that in the model outdoors and on the map, the orbits of the planets were represented *all in the same plane*, level with the ground. On the map, they were also represented as all in the same plane. Point out that this is similar to the real Solar System, because the orbits of the planets and almost all the asteroids form a disk, rather than a ball. The orbits do tilt a little from a perfectly flat disk, but not much. The objects in the Kuiper Belt tilt quite a bit. This includes Pluto. Mention that objects in the Oort Cloud are thought to form a ball shape rather than a disk.

TEACHER CONSIDERATIONS

SESSION 3.9 The Outer Solar System and Beyond

6. **Discuss the scale model's accuracies and inaccuracies.** Have students share what was accurate about the scale model of the Solar System. [Comparative sizes of the objects and distances between them were depicted accurately.] Then ask students to list things that were not accurate about the class model. [It is still much smaller than the real Solar System. The planets did not move or orbit the Sun.]

7. **Optional: Have students complete a writing assignment for use as an embedded assessment.** Refer to the Assessment Opportunity on page 123 for more details.

8. **Revisit Questions #1 and #2 on the Pre-unit 3 Questionnaire with students.** Ask students to think back to the questionnaire that they took at the beginning of the unit. Remind them that they were shown a diagram model of the Solar System and asked to list the model's accuracies and inaccuracies. Place the transparency of the Pre-unit 3 Questionnaire on the overhead projector or project it from the computer. Cover up all the questions on the questionnaire except for the diagram of the Solar System and Questions #1 and #2. Ask the class to consider the diagram again. What accuracies and inaccuracies would they point out now, after completing the Solar System scale model activity?

Where Does the Solar System End?—Student Reading

1. **Considering where the Solar System ends.** Tell the class that today they considered the distances of the orbits of the planets in the Solar System. Remind them how far away Neptune was—the farthest planet out. Ask students if they think the Solar System ends at any point. Tell them that they will be reading an article that considers this very question.

2. **Pass out copies of Where Does the Solar System End?—Student Reading.** Have students read silently for a few minutes or read the article together, out loud, as a class.

3. **Discuss the heliosphere.** After everyone has finished reading, ask students to explain what the heliosphere is. [The heliosphere is the boundary where the solar wind from the Sun hits the dust and gas around the Solar System.] Ask, "Why is the heliosphere important?" [The heliosphere protects the Solar System from high-energy particles coming from beyond the Solar System.] Ask, "Where is the heliosphere?" [The closest parts of the heliosphere are more than 90 times the distance between the Earth and the Sun, or 90 AUs away.]

TEACHER CONSIDERATIONS

QUESTIONNAIRE CONNECTION
Use this opportunity to review Questions #1 and #2 of the Pre-unit 3 Questionnaire with your students.

ASSESSMENT OPPORTUNITY
EMBEDDED ASSESSMENT: SCALE OF THE SOLAR SYSTEM WRITING ASSIGNMENT
Have students write a response to the following question:

> What did you learn from the scale model of the Solar System? Explain two things that you learned from the scale-model activities in Sessions 3.8 and 3.9. (The outdoor scale model of the inner Solar System and the mapped scale model of the outer Solar System.)

Students' written work can be used as an embedded assessment. See the scoring guide on page 123 in the Assessment section.

PROVIDING MORE EXPERIENCE
You may assign the reading as homework and give students copies of Where Does the Solar System End?—Reading Questions (from the copymaster packet or the CD-ROM file) to answer as a written homework assignment. This is a good option if you don't have time to go through the outlined class discussion of the reading.

SESSION 3.9 The Outer Solar System and Beyond

4. **Why might the heliosphere be a boundary for the Solar System?** Ask the class why studying the heliosphere might be a way for scientists to learn more about where the Solar System ends. [Unlike light or gravity, neither of which have defined boundaries, the heliosphere forms a "bubble" around the Solar System. It is an indication of how far the solar wind has permeated into the space around the Solar System.]

5. **For now, there is no definite answer.** Conclude the discussion by telling the class that scientists are still working to learn more about where the Solar System ends.

A Swirling Cloud Model

1. **The Solar System is an organized system.** Tell the class that the Solar System is an organized system, which means that it is arranged in a structured way. Ask students to describe how the Solar System is arranged. [All of the planets orbit the Sun—the center of the Solar System.] Ask the class if they can think of another organized system in the Solar System itself that they have studied. [Jupiter and its moons.] Ask how the Solar System and the Jupiter system are similar to one another. [Both systems consist of objects moving around one central object.]

2. **Teams discuss their ideas about how the Solar System formed.** Arrange students into teams of 4–6. Ask them to discuss how they think an organized system like the Solar System might have formed. Give teams a minute or two to discuss some of their ideas.

3. **Teams share their ideas.** Ask a few teams to share their ideas with the class.

4. **Using a model to study formation of the Solar System.** Tell the class that the scale model they completed today gave them a realistic sense of the various sizes and distances of objects in the Solar System. Tell them that now they will use another model to see how an organized system like the Solar System (with objects swirling around a central object) might have formed. Remind students of the key concept posted on the concept wall from Session 3.1:

   Scientists use models to demonstrate ideas, explain observations, and make predictions.

5. **Explain the model.** Tell students that they are going to use a container of water as a model to represent the region of space where the Solar System is now, but as it was a very, very long time ago *before* the Solar System existed.

TEACHER CONSIDERATIONS

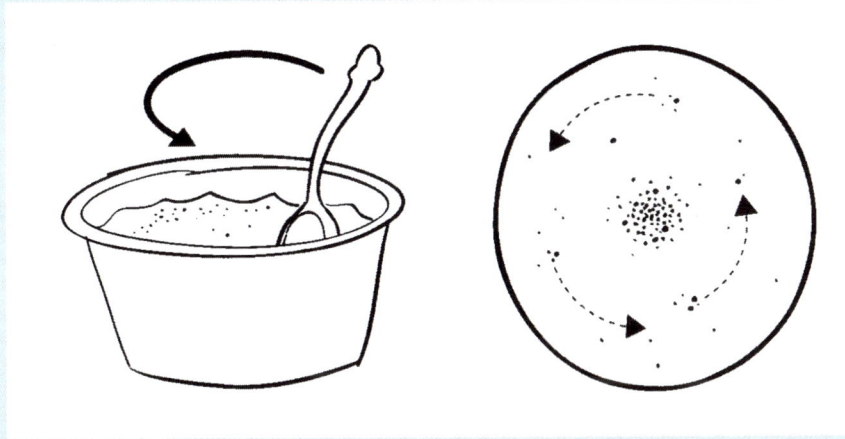

TEACHING NOTES

Be respectful and nonjudgmental of students in your class whose personal beliefs may include creationism. If any of your students raise this issue during class, address it briefly but tactfully. Let the class know that the swirling cloud theory is a scientific view of how the Solar System might have formed, based on evidence that scientists have been able to put together. It is still developing as scientists keep learning more. As a science teacher, you are sharing this scientific theory with the class. The *National Science Education Standards* states, "Explanations on how the natural world changed based on myths, personal beliefs, religious values, mystical inspiration, superstition, or authority may be personally useful and socially relevant, but they are not scientific." The National Science Teacher's Association statement on "The Nature of Science and Scientific Theories" quotes this passage, then adds, "Because science can only use natural explanations and not supernatural ones, science teachers should not advocate any religious view about creation, nor advocate the converse: that there is no possibility of supernatural influence in bringing about the universe as we know it."

See http://www.nsta.org/about/positions/evolution.aspx

SESSION 3.9 The Outer Solar System and Beyond

6. **The pepper represents dust and gas.** Sprinkle one or two shakes of pepper into each prepared container of water. Tell students that the pepper represents dust and gas in the region of space to be occupied by the Solar System. Point out that the pepper in the water is not organized in any particular way.

7. **Forming the Solar System using the model.** Explain that students will take turns using the model in their groups. Everyone should have a chance to try it. To work the model, they should stir the water in a circle a few times and then remove the stir stick and carefully observe what happens. Afterward, each student should prepare the model for the next person by using the stir stick to randomly redistribute the pepper in the water.

8. **Students should think about the following questions as they run the model:**
   - What did the pepper do?
   - How do you think the behavior of the pepper represents the formation of the Solar System?
   - What do you think is accurate about the model, and what do you think is not accurate?

9. **Distribute a container of peppered water and a stir stick to each team.** Let students run the model. Circulate among the teams, reminding them to remove the stir stick and watch carefully each time they run the model. It should take only a few minutes for all students to get a turn.

10. **Collect materials and clean up.** Have students dry the tables with paper towels or rags if there were any spills.

11. **Call on students to describe the behavior of the pepper.** Ask, "What did the pepper do?" [Most of the pepper forms a clump in the middle of the bottom of the container with just a few grains circling at various distances from the center.]

12. **Making sense of the behavior of the pepper.** Ask students what they think the clump of pepper represents. [The Sun.] Ask what they think the grains of pepper that circle at a distance from the clump represent. [Other objects in the Solar System that orbit the Sun, such as planets and asteroids.]

TEACHER CONSIDERATIONS

TEACHING NOTES
For a more accurate swirling cloud model, tell your students to stir the peppered water in a counterclockwise direction as seen from the north, since the Solar System also swirls counterclockwise.

SESSION 3.9 The Outer Solar System and Beyond

13. **Ask students what is accurate about the model.** Possible answers include:

    - Randomly sprinkled pepper forms into an organized system with some pepper clumped in the center, and the remaining pepper swirling around it. The Solar System, an organized and structured system, was formed from an initially disorganized system. (This is an important point—be sure to emphasize it!)
    - Most of the pepper ends up in the middle of the system. The Sun constitutes most of the matter in the Solar System and is located in the middle of the Solar System.
    - The pepper circles in the same direction. Almost all objects in the Solar System orbit in the same direction.
    - The pepper swirls in the same plane, forming a thin disk. In the Solar System, the orbits of planets and most asteroids are nearly in the same plane.

14. **Ask students what about the model is not accurate.** Possible answers include:

    - The pepper moves with the currents of water in the container, but the movement of objects in the Solar System is controlled primarily by gravity.
    - The swirling pepper eventually slows down and stops moving. The Solar System is continuously in motion.
    - All of the pepper orbits in the same plane, but some objects in the Solar System, such as comets and KBOs, do not.
    - The pepper and water model is very small and not to scale. The Solar System is considerably larger!

15. **Understanding the swirling cloud model.** Explain to students that the model they just worked with represents how scientists think the Solar System may have formed. Point out that, even with its inaccuracies, the model is useful for easily visualizing how a disordered system might have become a structured and ordered one. Post on the concept wall, under Key Space Science Concepts:

    The Solar System formed from gas and dust, which gravity drew together into a whirling system.

TEACHER CONSIDERATIONS

SCIENCE NOTES

How does the swirling cloud model work? In reality, organized systems like the Solar System and the Jupiter system are formed when gravity pulls matter together. In the swirling cloud model, the pepper simply follows the currents of swirling water. At the very bottom of the container, there is a slight current toward the middle. The pepper that sinks to the very bottom follows this current and forms a clump, while the pepper that stays suspended higher up orbits the center of the container. The reason for the flow of current toward the center at the bottom of the container is complex—students do not need to know why this occurs in order to appreciate the formation of an orderly system out of a chaotic one.

One teacher said, "I was surprised that students made such great observations about the comparisons between the pepper model and the Solar System. It's funny how an activity as simple as this can clearly illustrate how the Solar System might have formed. My kids enjoyed trying to change the pattern of the clump at the bottom of the dish."

SESSION 3.10

Human-powered Orrery

Overview

An orrery is a model of the Solar System that illustrates the relative motions and positions of bodies in the Solar System. The name comes from Charles Boyle, the 4th Earl of Orrery, for whom one of these models was made. The first orreries were mechanical, but a computer model of the Solar System, such as the one on the Space Science Sequence CD-ROM, is also called an orrery. In this session, the class works together to create a human-powered orrery to model the movements of the four inner planets. Students assist in setting up this moving model of the Solar System and take turns playing the roles of Mercury, Venus, Earth, and Mars. As the class observes the orrery in motion, they form conclusions about the orbital periods of the inner planets. Afterward, the class predicts the orbital periods of the outer planets using the mapped scale model transparency from Session 3.9. During this session, the key concepts that will be added to the classroom concept wall are:

- *Planets closer to the Sun have smaller orbits and move more quickly than planets farther from the Sun.*
- *Objects in the Solar System are in regular and predictable motion.*
- *As seen from Earth, the positions of the planets and the Sun are always changing.*

Human-powered Orrery	Estimated Time
Setting Up and Running the Orrery	35 minutes
Reflecting on Planetary Motion	10 minutes
Total	**45 minutes**

What You Need

For the class:
- ❑ overhead projector or computer with large-screen or LCD projector
- ❑ (optional) Space Science Sequence CD-ROM
- ❑ prepared key concept sheets from the copymaster packet or CD-ROM file
- ❑ transparency of the map with outer planet orbits drawn in (from Session 3.9)
- ❑ a 2.5 meter piece of thin rope or string (made out of a material that is not stretchy)
- ❑ 5 index cards (4"x6")
- ❑ marble, close to 1.4 cm in diameter
- ❑ several rolls of masking tape or blue painter's tape or chalk

Unit Goals

The Solar System is centered around the Sun, the only star in the Solar System.

A wide variety of objects orbit the Sun in the Solar System.

Scientists categorize Solar System objects according to their characteristics; however, not all objects can be easily categorized.

Objects in the Solar System are in regular and predictable motion.

The Solar System is mostly empty space, and is very large compared to the objects located within it.

TEACHER CONSIDERATIONS

TEACHING NOTES

More about the orrery. An orrery is a moving model of the Solar System that includes the Sun and planets and sometimes the moons of the planets. It is used to demonstrate the motions and positions of objects in the Solar System. Around 1704, one of the first orreries was designed by George Graham, an engineer and inventor. It was built by John Rowley, a maker of scientific instruments. The orrery was given as a gift to Charles Boyle, the 4th Earl of Orrery. It was the Earl whose name stuck to the device.

In a spatial sense, the human-powered orrery is a two-dimensional model. (Although the participants in the orrery are three-dimensional, they are representing the orbits of the planets on a flat, two-dimensional surface.) This model is, however, three-dimensional in the sense that it changes through time. Two dimensions of space and one dimension of time make up the three dimensions.

The Space Science Sequence CD-ROM has an interactive orrery of the outer Solar System (CLICK on the Moving Planets activity), which you may find useful in demonstrating orbital movement to your students. The animation may not be necessary for older or more advanced students.

Key Vocabulary

Scientific Inquiry Vocabulary

Category
Characteristic
Evidence
Model
Observation
Prediction
Scale
Scale model
Scientific explanation

Space Science Vocabulary

Asteroid
Astronomical Unit (AU)
Comet
Diameter
Heliosphere
Kuiper Belt Object (KBO)
Moon
Orbit
Planet
Sphere
Star
System

UNIT 3 • 503

SESSION 3.10 Human-powered Orrery

For each team of 4-6 students:
- ❏ scratch paper
- ❏ pencil

Getting Ready

1. **Prepare the key concept sheets.** Make a copy of each key concept and have them ready to post onto the classroom concept wall during the session.

2. **Find a clear space at least five meters (about 17 feet) square for the orrery.** It doesn't matter whether the space is located indoors or outdoors, as long as it has a smooth surface that can be marked with tape or chalk.

3. **Tie knots in the rope corresponding to the orbits of the inner planets.** Tie a large knot at one end of the rope. This will indicate the starting point of the model—the Sun. Measuring from this first knot out, tie four more knots at the following distances along the rope for each planet's orbit: 58 centimeters (Mercury), 108 centimeters (Venus), 150 centimeters (Earth), and 228 centimeters (Mars). (Note: These distances are all distances from the Sun's knot, NOT the distances between the knots themselves!)

4. **Have several rolls of masking tape, blue painter's tape, or chalk on hand.** The tape will be used to mark out the orbits of the planets. Choose tape that is at least one inch wide and a color that contrasts with the color of the surface on which the class will be working. If the model will be staged outdoors on a paved surface, chalk is an easy and convenient alternative to tape.

5. **Make signs for the Sun and inner planets.** On an index card, write "SUN" in large, bold lettering. Tape the marble to the sign. Use the other cards to make signs labeled "MERCURY," "VENUS," "EARTH," and "MARS." Put a tiny dot, much less than a millimeter across, on each sign to represent each planet.

The gym, cafeteria, multi-purpose room, music room, school lobby, or paved outdoor schoolyard are all possible location options for this activity.

These scaled planet orbit distances are one-hundredth of the distances used for the outdoor scale model in Session 3.8.

TEACHER CONSIDERATIONS

TEACHING NOTES

The key concepts can be posted in many different ways. If you don't want to use sentence sheets, here are some alternatives:

- Write the key concepts out on sentence strips.
- Write the key concepts out before class on a posted piece of butcher paper. Cover each concept with a strip of butcher paper and reveal each one as it is brought up in the class discussion.

Instructions for Setting Up the Orrery by Yourself

We highly recommend that you involve your students in the marking of the planetary orbits for this activity. It is easier and faster to set-up the orrery with many pairs of hands helping you out, and the process of marking the orbits is one your students may find educative. If, however, it is necessary for you to set up the orbits by yourself before class, read through the orrery activity and then follow the modified set-up as described below:

1. Tie knots in the rope corresponding to the orbits of the inner planets. (See Step #3 on page 504 under Getting Ready.) Tie a large knot at one end of the rope—this will indicate the starting point of the model, the Sun. Measuring from this first knot out, tie four more knots at the following distances along the rope for each planet's orbit: 58 cm (Mercury), 108 cm (Venus), 150 cm (Earth) and 228 cm (Mars). Note that these distances are all distances from the Sun's knot, NOT the distances between the knots themselves!

2. Prepare pieces of tape or have chalk ready to mark the orbits. If using tape, you will need 6 pieces for Mercury's orbit, 16 pieces for Venus' orbit, 26 pieces for Earth's orbit, and 50 pieces for Mars' orbit.

3. Cut a piece of cardboard (or poster board) into strips of varying lengths to help you determine the correct distance between tape pieces (or chalk marks) for each planet's orbit. For Mercury, you'll need a 61 cm cardboard strip; for Venus, a 42 cm strip; for Earth, a 36 cm strip; and for Mars, a 29 cm strip.

4. Mark out the orbit for each planet using the knotted rope, the cardboard strips and the pieces of tape (or chalk). For example, to mark out Mercury's orbit:

   - In the space you have designated for the orrery, make an X using the tape (or chalk) to represent the Sun's position.
   - Place the knotted end of the rope at the X, and carefully stretch the rope straight out to the first knot (tied 58 cm from the Sun's knot). Make sure that the knot at the end of the rope remains on the X. Place a piece of tape (or use the chalk) to mark the position of the first knot on the ground.
   - Place one end of the 61 cm cardboard strip at the first tape (or chalk) mark, and place a second tape (or chalk) mark on the

continued on page 507

SESSION 3.10 Human-powered Orrery

6. **Recruit additional people if necessary.** Setting up the orrery itself is an integral part of this session and requires the participation of at least 26 people. If you don't have that many students, try to find additional people such as other faculty members, teacher's aides, or students from other classes to help with this activity.

7. **Decide how you will divide the class into teams of 4–6 students for the Reflecting on Planetary Motion discussion.**

8. **Optional: If you are planning to show the CD-ROM orrery animation (in the Moving Planets activity), set up a computer with a large-screen monitor or LCD projector.**

GO! Setting Up and Running the Orrery

1. **Remind students of the scale models from Sessions 3.8 and 3.9.** Ask students to think back to the outdoor scale model of the inner planets and the mapped scale model of the outer planets. Have students describe their impressions of the scale of distances in the Solar System. [Uncrowded, mostly space, pretty empty.] Point out that these scale models allowed the class to visualize the vast distances between the planets of the Solar System.

2. **Tell students that today they will construct another scale model to observe the movements of the planets more closely.** Remind students that at the end of the last session, they observed the swirling cloud model and saw how an ordered system could be formed from a disordered one. Tell them that today they will observe another moving model of the Solar System, called an *orrery*. This model will allow them to look more closely at the movements of the planets around the Sun.

3. **The scale of the orrery.** Today's model will be *one-hundred billionth* the size of the actual Solar System. Remind students that the outdoor scale model was one-billionth the size of the Solar System, so this model is one-hundredth the size of that model.

4. **Explain the scale of the Sun and planets in this model.** Show students the sign that says "SUN" with the marble taped to it. Tell them that the diameter of the marble is about one one-hundred billionth the diameter of the Sun. The size of the Sun and the sizes of the planets' orbits *will be* to scale. Show the signs for the four inner planets. The planets will be represented by tiny specks, which *are not* to scale. They are too large! Explain to the class that the focus of today's model is on the *movement* of the planets and not the sizes of the planets, so scaled sizes are not as important.

TEACHER CONSIDERATIONS

Instructions for Setting Up the Orrery by Yourself
continued from page 505

ground at the other end of the 61 cm cardboard strip. Continue using the cardboard strip to mark out distances between the tape (or chalk) marks until you have used 6 tape (or chalk) marks total outlining Mercury's orbit. (The pieces of tape or chalk marks should should appear evenly spaced apart.) While marking out the orbit, you should occasionally use the knotted rope to make sure that the tape (or chalk) marks are being placed 58 cm away from the "Sun".

Repeat the above procedure for the remaining three planets, using the corresponding rope knot and cardboard strip length to correctly mark out each planet's orbit.

TEACHING NOTES

A note about Setting Up and Running the Orrery: This is a particularly detailed activity. In order to make the presentation write-up easier for you to follow, additional section sub-headings have been added:

- Modeling Mercury's Movement
- Comparing the Movements of Mercury and Venus
- Setting the Orbits of Earth and Mars
- Running the Model of the Inner Solar System

Note that these subheadings are **not** listed in the session timetable. The 35 minutes allotted for Setting Up and Running the Orrery includes them.

Running the orrery may pose some class-management challenges. Listed below are a few suggestions and ideas that may address some of the most common behavioral issues with this activity.

Maintaining student interest. Involve your students in all aspects of the orrery. When describing a student's role in the orrery, address the entire class, not just the student. Give everyone in the class a chance to participate by constantly changing student roles—allow several different students the chance to play the role of the Sun or one of the moving planets. Clapping and chanting "two weeks" in unison helps to center and focus the group. During the activity, pose observation questions to the students who are not involved in the orrery directly and make a mental note to swap those students into the moving orrery at some point.

Managing the class. Be prepared to deal with class-management issues during this activity. If you know that this will be a significant issue, consider showing the class the orrery animation on the Space Science Sequence CD-ROM (CLICK on the Moving Planets activity) instead of having them run the orrery themselves.

SESSION 3.10 Human-powered Orrery

One teacher said, "I kept having the planets race around the Sun, and despite numerous attempts, Mercury kept winning. Thanks for taking a really tough abstract concept and making it easy for the kids to understand."

5. **They will all play important roles in the orrery.** Tell the class that today's activity is called the Human-powered Orrery, and that everyone will be an integral and important part of the moving model!

Modeling Mercury's Movement

1. **Move the class to the space chosen for the orrery.** Take the knotted rope, tape or chalk, and cardboard signs with you.

2. **Have students stand in a circle.** In the center of the circle, mark an X with the tape or chalk. Tell the class that the X represents the position of the Sun. Select a student to be the Sun and have him or her stand on the X in the middle of the circle with the "SUN" sign. Remind the class that in this model the sizes of the objects are not to scale. Tell the students that if the Sun were to scale, it would be less than 15 millimeters (13.9 millimeters) across in this model.

3. **Select six students to help form the orbit of Mercury.** Assure the class that everyone will have a chance to participate in the model at some point. Give each Mercury student a short piece of tape (about 4" long) or a piece of chalk and have them form a circle around the Sun.

4. **Explain to the class how each planet's orbit will be set.** Tell the class that since today's scale model is only one-hundredth the size of the previous scale model, the distances of the planets from the Sun will also be one-hundredth the distances that they were in the other scale model. Show the class the length of knotted rope and tell them that each of the knots on the rope represents the scaled distance of a planet from the Sun. The rope will be used to set the orbits of the four inner planets.

5. **Set the size of the orbit of Mercury.** Have the Sun student hold onto the knot at the end of the rope. Hold onto the first knot (tied 58 centimeters from the Sun's knot) and walk in a circle around the Sun, making sure to pull the rope taut as you delineate Mercury's orbit. The Sun student should turn with you as you walk around. Ask the six Mercury students to space themselves evenly around the Sun along the orbit's path. Once all six are in place, have them each mark their position with their tape or chalk, and then have them rejoin the larger class circle.

TEACHER CONSIDERATIONS

HUMAN-POWERED ORRERY

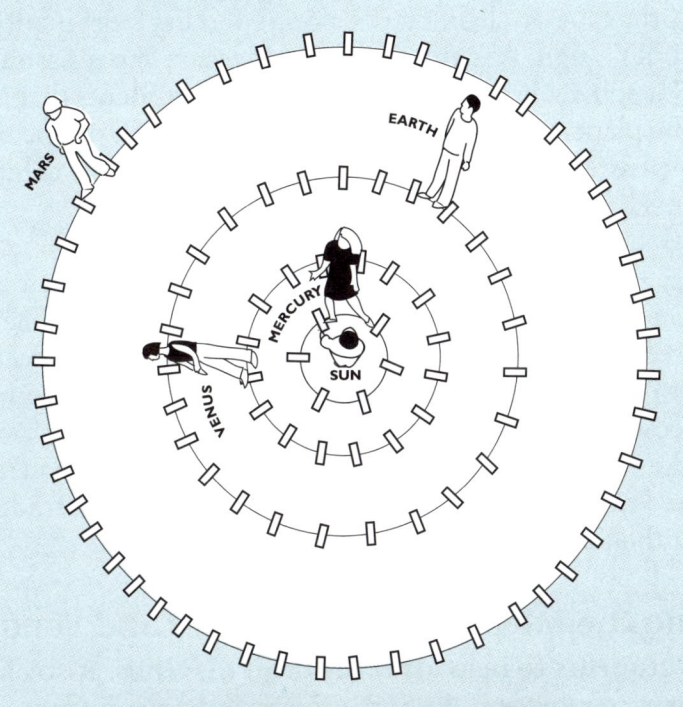

Mercury: 58 cm from Sun / 6 pieces of tape
Venus: 108 cm from Sun / 16 pieces of tape
Earth: 150 cm from Sun / 26 pieces of tape
Mars: 228 cm from Sun / 50 pieces of tape

TEACHING NOTES

Give as many students as possible an opportunity to participate in the orrery. Choose one student to be the Sun for now, but swap in other students for the Sun's role throughout the activity.

Consider allowing different students to set the orbits of the planets using the knotted rope. This option allows more students to participate in the orrery and can help maintain student engagement in the activity. You may first want to demonstrate to the class by setting Mercury's orbit and then allowing students to set the orbits of the other inner planets.

SESSION 3.10 Human-powered Orrery

6. **Choose a student to represent Mercury.** Have the student stand on one of the tape or chalk marks along Mercury's orbit with the "MERCURY" sign. Ask the class how Mercury moves around the Sun. [It spins while orbiting the Sun.] Tell students that even though the planets do spin around an axis, they will not be modeling this today. Instead, today's model will focus on just the orbital movements of the planets.

7. **Modeling Mercury's orbital movement.** Tell the class that just as the model has a distance scale, it also has a time scale. The time represented between each tape mark is about two Earth weeks. Have the Mercury student step from tape mark to tape mark around the Sun in a counterclockwise orbit. Ask the class how many weeks it would take Mercury to make a full orbit around the Sun using this time scale. [12 Earth weeks.] Ask the Sun and Mercury students to hand you their signs and rejoin the larger class circle.

Comparing the Movements of Mercury and Venus

1. **Select 16 students to help form the orbit of Venus.** Also choose a new student to represent the Sun. Give each Venus student a short piece of tape or a piece of chalk and have them form a circle around the Sun that is larger than Mercury's orbit.

2. **Set the size of the orbit of Venus.** Again, have the Sun student hold onto the knot at the end of the rope. Hold onto the second knot (tied 108 centimeters from the Sun's knot) and walk in a circle around the Sun, making sure to pull the rope taut as you delineate Venus' orbit. Again, the Sun student should turn with you as you walk around. Ask the 16 Venus students to space themselves evenly around the Sun along the orbit's path. Once all 16 students are in place, have them each mark their position with their tape or chalk and then have them rejoin the larger class circle.

3. **Students compare Mercury's tape marks to Venus' tape marks.** Have students describe how the tape marks are spaced. [Venus' tape marks are closer together than Mercury's tape marks.] Ask the class how much time is represented between each tape mark. [Two Earth weeks.] Ask the class which planet, Mercury or Venus, will move a greater distance in a two-week time period. [Mercury.]

TEACHER CONSIDERATIONS

TEACHING NOTES

You may want to explain how unusually Mercury moves. Mercury spins completely around three times for every two orbits it makes around the Sun.

About the terms *rotation* and *revolution*. Traditionally, students have been taught to use the word *rotation* to describe the spinning of a planet and the word *revolution* to describe the planet's movement around the Sun. For example: "The Earth *rotates* on its axis and *revolves* around the Sun." In other contexts, the words *revolve* and *rotate* can take on slightly different and confusing meanings. *Revolve* often means "to spin," as in a "revolving door." *Rotate* can be used to mean "to follow a cyclical path or system," as when a certain task or responsibility is rotated among several people. The terms *spin* and *orbit* are scientifically accurate alternatives that are clearer to students.

SESSION 3.10 Human-powered Orrery

Having the class clap and announce the time scale together is an excellent way to focus all students on the activity at hand.

4. **Run the model with Mercury and Venus.** Choose two students to represent Mercury and Venus and pass out the signs. Explain that both planets must move around the Sun according to the same time scale. To help the two planets synchronize their movements, the class will clap and announce "two weeks." With each clap, Mercury and Venus should move along their orbits from one tape mark to the next, counterclockwise around the Sun. Start off slowly at first, clapping about once every two seconds. Pick up the pace after a couple of claps.

5. **Comparing Mercury and Venus.** After a dozen or so claps, stop the class and ask, "If Mercury and Venus were racing around the Sun, who do you think would win the race?" [Mercury.] Have the students who are representing the Sun, Mercury, and Venus hand you their signs and rejoin the larger class circle.

Setting the Orbits of Earth and Mars

1. **Select 26 students to help form the orbit of Earth.** Also, choose a new student to represent the Sun. Give each Earth student a short piece of tape or a piece of chalk and have them form a circle around the Sun that is larger than Venus' orbit.

2. **Set the size of the orbit of Earth.** Again, have the Sun student hold onto the knot at the end of the rope. Hold onto the third knot (tied 150 centimeters from the Sun's knot) and walk in a circle around the Sun, making sure to pull the rope taut as you delineate Earth's orbit. The Sun student should turn with you as you walk around. Ask the 26 Earth students to space themselves evenly around the Sun, along the orbit's path. Once all 26 students are in place, have them each mark their position with their tape or chalk and then rejoin the larger class circle.

3. **Have students compare orbits of Mercury, Venus, and Earth.** Students should note that the tape marks in Earth's orbit are closer together than the tape marks in Venus' orbit. Ask students whether Earth will move slower or faster than Venus. [Slower.] Ask students to predict whether Mars will move faster or slower than Earth.

TEACHER CONSIDERATIONS

SESSION 3.10 Human-powered Orrery

This makes a circle of 50 marked steps for the orbit of Mars. (In reality, 49 tape marks would be more accurate, but it's easier to set the orbit of Mars with an even number of marked steps.)

4. **Select 25 students to help form the orbit of Mars.** Give each Mars student *two* pieces of tape or a piece of chalk. Just as before, have the Sun hold onto the knot at the end of the rope while you walk out a 228-centimeter orbit around the Sun. Ask the 25 Mars students to space themselves evenly around the Sun, along the orbit's path. Once all 25 students are in place, have them each mark their position with tape or chalk. Then they should each shift half a space to their right and mark another position halfway between the marks on either side. Have the students rejoin the larger class circle.

Running the Model of the Inner Solar System

1. **Select students to represent Mercury, Venus, Earth, and Mars.** Pass out the signs and have the students representing the four inner planets all step to a marked position in their orbit on the same side of the Sun. The planets should be lined up as if they are about to begin a race. Emphasize that this is an unusual planetary alignment. Have students make predictions about the rates of movement of the four inner planets. Ask, "If the planets were racing to complete their orbits around the Sun, which planet would you want to be if you wanted to win this race?" [Mercury.]

2. **Run the model.** Have everyone begin clapping together. Lead the class in chanting "two weeks" with each clap. Stop the class after 26 claps. Point out that Earth has made one full orbit around the Sun. Ask the class how many weeks have passed. [52 weeks, or one year.] Have students describe the progress of the other planets and how much time has gone by for them.

3. **If time allows, continue running the model with different students.** Try to make sure that everyone in the class gets a chance to participate in the orrery. Solicit comments and observations from students as they observe the model in action. Return to the classroom 10–15 minutes before the period ends.

Reflecting on Planetary Motion

1. **Divide students into teams of 4–6.** Give each team a pencil and a piece of scratch paper. Ask them to list any "true statements" they can make after observing the movements of the planets in the orrery. After a few minutes, call on each team to read one of its statements aloud.

2. **Different planets require different amounts of time to complete one orbit around the Sun.** Ask teams to discuss whether it is the length of the planet's orbit or the speed at which the planet moves that makes the difference in the time it takes to make a full orbit. [Both the speed of the planet and the length of its orbit contribute.]

TEACHER CONSIDERATIONS

If you used strips of tape for marking the orbits, ask students to help remove the tape from the ground. Alternatively, if you are planning to run the orrery with more than one class, you could leave the tape marks where they are. (This option would not allow the next class to participate in the orrery setup.)

SESSION 3.10 Human-powered Orrery

3. **Considering the *year lengths* of the inner planets.** Ask students what the term *year length* might mean. [A year is defined as the amount of time it takes for a planet to complete its orbit around the Sun.] Ask students to think back to the orrery and the year lengths of the four inner planets. Ask, "Which planet has the shortest year length?" "Which planet has the longest year length?" [Mercury has the shortest. Mars has the longest.]

4. **Looking at the outer planets.** Display the overhead transparency from Session 3.9 that shows the mapped orbits of the outer planets. Ask the class why the orrery included only the orbits of the inner planets. [The orbits of the outer planets are spread out even more than those of the inner planets. Even scaled down to one-hundred billionth of their real size, the orbits of the outer planets would not fit within the schoolyard.]

5. **Discussing the *year lengths* of the outer planets.** Ask students what they might predict about the year lengths of the outer planets. [They are very long—longer than the year lengths of the inner planets.] Read off the year lengths of the planets to the class:

    Mercury's year = 12 Earth weeks or 0.2 Earth years
    Venus' year = 32 Earth weeks or 0.6 Earth years
    Mars' year = 2 Earth years
    Jupiter's year = 12 Earth years
    Saturn's year = 29 Earth years
    Uranus' year = 84 Earth years
    Neptune's year = 165 Earth years

6. **Optional: Show the class the orrery animation (click on the Moving Planets activity) on the Space Science Sequence CD-ROM.**

7. **A pattern to the year lengths.** Ask students if they notice a pattern or trend between the location of the planets and their year lengths. [The farther away from the Sun a planet is, the longer its year length.] Ask students why they think that planets closer to the Sun have smaller year lengths. [These planets have smaller orbits and they move more quickly than the planets farther out from the Sun.] Post on the concept wall, under Key Space Science Concepts:

    Planets closer to the Sun have smaller orbits and move more quickly than planets farther from the Sun.

TEACHER CONSIDERATIONS

CD-ROM NOTES
SESSION 3.10: ORRERY ACTIVITY OF THE OUTER SOLAR SYSTEM

Use this interactive animation to simulate the class orrery with the outer planets of the Solar System. Each colored robot represents one of the outer planets. The user has three options in running this orrery:

1) Have the planets move along their orbits in increments of two weeks.
2) Have the planets move along their orbits in increments of one Earth year.
3) Have the planets move along their orbits continuously.

Elapsed time as counted in Earth weeks is displayed in the lower right corner of the screen. To enlarge the interactive to full screen, press CONTROL F (Windows) or APPLE F (Macs). Press ESC to exit the full-screen display.

SESSION 3.10 Human-powered Orrery

8. **Viewing the movements of the planets from the Sun.** Ask the class how the movements of the planets would appear to someone standing in the position of the Sun. [The planets would appear to move in an organized and orderly manner around the Sun. The viewer would see a consistent pattern in the motion of the planets.] Post on the concept wall, under Key Space Science Concepts:

   Objects in the Solar System are in regular and predictable motion.

9. **Viewing the movements of the planets from the Earth.** Now ask the class to visualize how the movements of the planets might appear to someone standing in the position of the Earth. [Since Earth is not in the center of the Solar System and is also constantly moving itself, the movements of the planets would seem complex to an observer on Earth.] Post on the concept wall, under Key Space Science Concepts:

   As seen from Earth, the positions of the planets and the Sun are always changing.

10. **Conclude by having teams list the accuracies and inaccuracies of the orrery.** Some things the model showed accurately:

    - All the planets orbited the Sun in the same direction. (Use this opportunity to remind the class of the swirling cloud model from Session 3.9.)
    - All the planets' orbits were in the same plane. (This is roughly true in the Solar System. The planes in which each planet orbits are different from each other by a few degrees.)
    - The orbits were all close to circular. (The orbits of all the Solar System's planets are close to circular, but each is slightly elliptical. The orbit of Mercury is noticeably more elliptical than the other orbits.)
    - The inner planets moved faster and had shorter orbits than the outer planets.

    Some things the model showed inaccurately:

    - The sizes of the planets were not to scale.
    - The planets did not spin.

TEACHER CONSIDERATIONS

PROVIDING MORE EXPERIENCE
If your students are able to quickly grasp the essential points of this model, and you wish to have a more extensive discussion, here are some additional and more subtle topics you can discuss with the class:

- **Orientation of the Sun, Earth, and Mars.** About every two Earth years, the Earth passes between the Sun and Mars. When this occurs, Mars appears high in the sky at midnight. If the Sun is between the Earth and Mars, Mars will not be visible at any time of day from Earth. Ask your students to try and explain these two scenarios.
- **Rising and setting of Mercury and Venus.** Earth will never pass between the Sun and Mercury or between the Sun and Venus. As a result, these planets are never visible from Earth at midnight. During an Earth day, Mercury and Venus rise and set fairly close to the time that the Sun rises and sets. These two planets are visible in a dark sky only shortly before sunrise or shortly after sunset. (Venus is often called the "evening star" or the "morning star.") Ask your students to try to explain these observations about Mercury and Venus.
- **Minimizing the traveling distance to Mars.** Ask students to think about how scientists planning a trip to Mars might minimize the distance required to make the trip.
- **Conjunctions.** When the Earth, Sun, and another planet are all lined up, the other planet is said to be in conjunction. If the planet is on the opposite side of the Sun from Earth, then the conjunction is called a superior conjunction. If the planet is on the same side of the Sun as Earth, then the conjunction is called an inferior conjunction.
- **Transits.** When a planet is in inferior conjunction and is nearly perfectly in line with the Earth and Sun, it appears right in front of the Sun and looks like a tiny black spot on the Sun. This is called a transit of the Sun by that planet. Only two planets can ever transit the Sun. Ask your students which two planets can transit the Sun.

SESSION 3.11

Evidence Circles and the Post-unit 3 Questionnaire

Overview

Evidence circles encourage students to think about and discuss an issue as scientists. As students attempt to support their point of view with evidence-based arguments, they must also assimilate all they have learned throughout the unit about the Solar System and many of its objects. In this session, students work in evidence circles to discuss the question of whether or not Pluto is a planet. Although this is a real question circulating among astronomers, the focus of the evidence circles is not *primarily* about Pluto. It is more a question about whether or not and/or how the category and definition of the term *planet* is useful in organizing objects in the Solar System. This activity gives students the opportunity to synthesize much of the information and concepts that are a part of this unit. The session closes with students filling out the Post-unit 3 Questionnaire. Both the questionnaire and students' written work about categorizing Pluto can be used to assess their learning in the unit. There are no key concepts for this session.

Evidence Circles and the Post-unit 3 Questionnaire	Estimated Time
Reviewing Main Unit Concepts	5 minutes
Evidence Circles: Is Pluto a Planet?	20 minutes
Taking the Post-unit 3 Questionnaire	15 minutes
Discussing the Questionnaire	5 minutes
Total	**45 minutes**

What You Need

For the class:
- ❏ the class list of What Makes an Object a Planet? (from Session 3.4)

For each team of 4–6 students:
- ❏ 1 copy of the Key for the Solar System Cards from Session 3.3
- ❏ (optional) 1 set of 36 Solar System cards from Session 3.3

For each student:
- ❏ 1 copy of the Evidence Circles: Is Pluto a Planet? student sheet from the copymaster packet or the CD-ROM file
- ❏ 1 copy of the Post-unit 3 Questionnaire (three pages) from the copymaster packet or the CD-ROM file
- ❏ a pencil

Getting Ready

1. Decide how you will divide the class into teams of 4–6 students for the Evidence Circles activity.

Unit Goals

The Solar System is centered around the Sun, the only star in the Solar System.

A wide variety of objects orbit the Sun in the Solar System.

Scientists categorize Solar System objects according to their characteristics; however, not all objects can be easily categorized.

Objects in the Solar System are in regular and predictable motion.

The Solar System is mostly empty space, and is very large compared to the objects located within it.

TEACHER CONSIDERATIONS

PROVIDING MORE EXPERIENCE

There are a large number of excellent sites on the Internet about the Solar System and related subjects. Here are some of the most helpful. Also see the Resources and References section on page 57.

A large collection of Astronomy related URLs
http://www.lhs.berkeley.edu/sii/URLs/URLs-Astronomy.html

The SETI Institute
http://www.seti.org

SETI@home
http://setiathome.ssl.berkeley.edu

NASA Educational Resources
http://education.nasa.gov/
http://nssdc.gsfc.nasa.gov/planetary/planetfact.html

Goddard Space Flight Center Space Science Education Home Page
http://www.gsfc.nasa.gov/education/education_home.html

Remote Sensing Public Access Center
http://www.rspac.ivv.nasa.gov

Hubble Space Telescope and Amazing Space (web-based educational activities)
http://www.stsci.edu/public.html

Astronomical Society of the Pacific
http://www.aspsky.org

The Nine Planets
http://www.seds.org/billa/tnp/nineplanets.html

The Nine Planets for Kids
http://www.tcsn.net/afiner/

PlanetScapes
http://planetscapes.com/

Earth-Moon System to Scale
http://www.dnr.state.oh.us/odnr/geo_survey/edu/hands10.htm

Global Quest: The Internet in the Classroom
http://quest.arc.nasa.gov

Smithsonian Institute "Museum Without Walls"
http://www.si.edu

It is up to you whether you would like your students to have access to the Solar System card sets during the Evidence Circle activity. Note that allowing students to use the cards will usually increase the amount of time needed for the activity.

Key Vocabulary

Scientific Inquiry Vocabulary

Category
Characteristic
Evidence
Model
Observation
Prediction
Scale
Scale model
Scientific explanation

Space Science Vocabulary

Asteroid
Astronomical Unit (AU)
Comet
Diameter
Heliosphere
Kuiper Belt Object (KBO)
Moon
Orbit
Planet
Sphere
Star
System

SESSION 3.11 Evidence Circles and the Post-unit 3 Questionnaire

2. **Make a copy of the Key for the Solar System Cards for each team.** You may also choose to have a set of the Solar System Cards available for each team.

3. **Make photocopies of the Evidence Circles: Is Pluto a Planet? student sheet and the Post-unit 3 Questionnaire (three pages).** Each student should have his or her own copy of each.

4. **Post the class list of What Makes an Object a Planet? from Session 3.4.**

GO! Reviewing Main Unit Concepts

1. **Highlight key concepts from this unit.** Have students look over the concept wall. Give them a minute to reflect on all they have learned about the Solar System in this unit. Direct their attention, in particular, to the following important key concepts:

   Many diverse objects make up the Solar System. (from Session 3.3)
   The Solar System is mostly empty space and is very large compared to the objects within it. (from Session 3.9)
   Objects in the Solar System are in regular and predictable motion. (from Session 3.10)

2. **Refer to the class list of What Makes an Object a Planet?**

   - It is large.
   - It is spherical.
   - It orbits the Sun.

   Tell students that this list is not complete. Scientists have considered other characteristics a planet must have. Ask the class whether they can think of other things they would want to know about an object in order to decide whether it belongs in the planet category. [What kind of orbit it has, what it is made of, what other objects it is similar to, etc.]

3. **Remind students that not all objects are easily categorized.** Refer to the posted key concept from Session 3.4:

   Not every Solar System object can be easily categorized.

   Tell the class that today they will discuss, in evidence circles, whether or not Pluto should be considered one of the planets. Students will use all they have learned from this unit to support their point of view.

TEACHER CONSIDERATIONS

EVIDENCE-CIRCLE RESPONSES: IS PLUTO A PLANET?

Use the students' answers to the questions on the Evidence Circles: Is Pluto a Planet? student sheet (especially Question #3) to gain insight into whether students:

- can categorize an object in the Solar System by its characteristics
- can apply factual information about the characteristics of an object to an argument about whether or not the object is a planet
- are able to use factual evidence in their written answers to back up their arguments

1. What makes an object a planet?
Answers should include:
   - It is round.
   - It is large.
   - It orbits the Sun.

2. Look over the Key for the Solar System Cards with your team. Discuss whether or not the two objects below are planets. For each object, write yes or no and then use facts as evidence to back up your opinion:

   a. Is Titan (#14) a planet?

   Arguments for no: It orbits Saturn, so it's a moon.

   Arguments for yes: It is spherical and large.

   b. Is Ceres (#17) a planet?

   Arguments for no: It's not large enough.

   Arguments for yes: It is spherical, fairly large, and orbits the Sun.

3. Is Pluto (#26) a planet? Write yes or no and then use facts to back up your opinion. Compare Pluto to one or both of the other objects listed in Question #2 to make your point.

   Arguments for no: It's too small—some moons and Eris are bigger than Pluto. Its own moon is almost as big as it is. It's almost as far away from the Sun as Quaoar, which is a KBO.

   Arguments for yes: It orbits the Sun and is pretty large and spherical, and has its own moon.

SESSION 3.11 Evidence Circles and the Post-unit 3 Questionnaire

You may want to have students note on their sheets which objects they disagreed about and why.

Evidence Circles: Is Pluto a Planet?

1. **Teams discuss the question: "Is Pluto a Planet?"** Divide the class into teams of about four students each. Pass out an Evidence Circles: Is Pluto a Planet? student sheet to each student. Go over the sheet with the class and tell students they are going to discuss the questions with their team. Each member should participate and contribute to the discussion. Like real scientists debating an issue, they may not agree on all points. Emphasize that this is okay!

2. **Scientists use evidence to back up arguments.** Tell teams that for Question #2 on the student sheet, they should use facts about each object to decide whether or not it fits the class list of what makes an object a planet.

3. **Pass out a Key for the Solar System Cards sheet to each team.** Tell students that they can look up facts about the sizes and orbits of the objects on this sheet.

4. **Optional: If you have decided to let your students use the Solar System card sets during this activity, pass them out to each team now.**

5. **Have groups discuss and fill out the Evidence Circles: Is Pluto a Planet? sheet.** Circulate among the groups. Respond to questions and make sure all students are participating. After 10–15 minutes, announce that they should begin finishing up their discussions and filling out their student sheets. Each student should answer whether or not the listed objects under Question #2 are planets and support his or her answer with sufficient evidence.

6. **Have teams present their case.** Take a few minutes to have some of the teams share their opinions and evidence. Take a hand-count survey to see how many students think Pluto should still be called a planet. Members of the same team do not necessarily have to agree with one another.

7. **Tell them that some scientists are still discussing this question.** Tell the class that despite Pluto's official loss of planetary status, some astronomers still feel that Pluto should be considered a planet. In general, however, astronomers tend to agree with the heartfelt statement: "Whether it is a planet or not, Pluto is still Pluto!"

TEACHER CONSIDERATIONS

ASSESSMENT OPPORTUNITY:
EMBEDDED ASSESSMENT: EVIDENCE-CIRCLE RESPONSES: IS PLUTO A PLANET?

Student responses on the Evidence Circles: Is Pluto a Planet? student sheet can be used as an embedded assessment. See the scoring guide on page 124 in the Assessment section.

TEACHING NOTES

Evidence circles. Is this your students' first experience with evidence circles? If so, spend some extra time introducing the procedure as follows:

What is evidence? Say that in this unit, the class has gathered lots of evidence about objects in the Solar System. Tell students (or remind them of) the definition of *scientific evidence*.

Define *scientific evidence*. Scientific evidence includes information or data gained from observations and investigations of nature and natural phenomena that can be used in making explanations.

Scientists discuss evidence from their investigations. Let students know that scientists rely on evidence to support what they believe to be accurate or true. They try to answer the following question: "What explanation best matches all the available evidence?"

The procedure for evidence circles. Tell the class that they are going to be discussing a question in small groups called evidence circles. As scientists, they will listen to each other, ask questions, and try to agree. They'll discuss whether or not each of the three objects on their student sheet is a planet.

Explain the procedure:
1. For each object, one student says what he/she thinks and the reasons why.
2. Other students who agree, add their reasons.
3. Then each student who disagrees says why and presents his or her reasons.
4. The group discusses with each other to see if they can come to agreement. If no one disagrees, they can talk about all the evidence that makes them convinced of their view.

continued on page 527

One teacher said, "I think Evidence Circles are a strategy I want to use in other lessons I teach. I heard students saying things like, 'But that's your opinion, it's not evidence.'"

Another said, "We had some excellent discussions about whether Pluto should be classified as a planet or not. The general consensus after our discussions was that we need a separate category for Pluto and Charon."

SESSION 3.11 Evidence Circles and the Post-unit 3 Questionnaire

Taking the Post-unit 3 Questionnaire

1. Collect the Evidence Circles: Is Pluto a Planet? student sheets and the Key for the Solar System Cards from each team.

2. Hand out the Post-unit 3 Questionnaires (three pages).

3. Give students about 15 minutes to fill out the questionnaire. Save five minutes at the end of the period to briefly discuss the questionnaires.

4. Collect the Post-unit 3 Questionnaires.

Discussing the Questionnaire

What have students learned? After collecting all of the questionnaires, take a few minutes to ask your students what they've learned from this unit. Follow up by asking, "Which parts of the questionnaire do you now feel you know much more about?"

TEACHER CONSIDERATIONS

TEACHING NOTES
continued from page 525

Scientists are open to changing their minds based on evidence.
The main point of evidence circles is for students to think about and discuss ideas and evidence in order to find the best explanation for something. Scientists are able to listen to the reasoning of others and to change their minds if they think their viewpoint is no longer supported by the evidence or if another explanation is better supported by the evidence. At the same time, scientists are not easily swayed by arguments that are not based on evidence. They must decide for themselves if a particular bit of evidence supports or does not support a particular explanation or position.

One teacher said, "This unit worked very well for my English Language Learners and special education inclusions. Getting away from standardized tests for assessment is great, likewise the journal writing opportunities."